ELOPMENTS

BIODEGRADATION OF CELLULOSE FIBERS

BACTERIOLOGY RESEARCH DEVELOPMENTS

Additional books in this series can be found on Nova's website at:

https://www.novapublishers.com/catalog/index.php?cPath=23_29&seriesp=
Bacteriology+Research+Developments

Additional E-books in this series can be found on Nova's website at:

https://www.novapublishers.com/catalog/index.php?cPath=23_29&seriespe=
Bacteriology+Research+Developments

BIODEGRADATION OF CELLULOSE FIBERS

BARBARA SIMONČIČ,
BRIGITA TOMŠIČ,
BORIS OREL,
AND
IVAN JERMAN

Nova Science Publishers, Inc.
New York

LIBRARY OF CONGRESS CATALOGING-IN-PUBLICATION DATA
Biodegradation of cellulose fibers / Barbara Simoncic ... [et al.].
p. cm.
Includes index.
ISBN 978-1-61668-154-8 (softcover)
1. Cellulose fibers. 2. Cellulose--Microbiology. 3. Cellulose--Biodegradation. 4. Microbial biotechnology. I. Simoncic, Barbara.
QR160.B565 2009
572'.56682--dc22
2009050560

Published by Nova Science Publishers, Inc. ✦ *New York*

CONTENTS

ABSTRACT

Natural cellulose fibers are high-molecular polymers that are strongly susceptible to microbial degradation. Fungal and bacterial attack on the fibers cause depolymerization of cellulose macromolecules, which is reflected in decreased molecular weight and strength, increased solubility and a changed crystallinity. Although biodegradable textile fibers are classified as environmentally-friendly materials, the biodegradation process that occurs when the textile product is still in use could cause serious functional, aesthetic and hygienic problems because of textile deterioration, staining, discoloration and odor. To protect the cellulose against biodegradation, chemical modification is of great importance. In this chapter, therefore, fibers were chemically modified by the use of an antimicrobial finish on the basis of AgCl (AG) in combination with a reactive organic-inorganic binder (RB), water and oil repellent finishes on the basis of fluoroalkylfunctional siloxane (FAS), perfluorooctyltriethoxysilane (PFOTES) in combination with di-ureapropyltriethoxysilane [bis(aminopropyl)terminated-polydimethylsiloxane (1000) (PDMSU) and aminopropyl-perfluoroisooctyl polyhedral oligomeric silsesquioxane ($AP_2PF_2IO_4$ POSS), as well as with an easy-care and durable press finish on the basis of imidasolidinone (DMeDHEU). Moreover, it was expected that these finishes would provide the active or passive antimicrobial properties of the modified fibers. Biodegradation of the finished cellulose fibers was carried out by the soil burial test according to SIST EN ISO 11721-2:2003. The chemical and functional properties of the fibers were determined by means of DP, SEM, AFM, FT-IR, XPS, ICP-MS, appropriate microbiological tests and static contact angle measurements of different liquids on the fiber surface. It can be seen from the study that AgCl embedded in RB imparts active antibacterial and antifungal properties to the coated fibers resulting in excellent microbial reduction and, consequently, strong inhibition of biodegradation. The presence of FAS, PFOTES-PDMSU and $AP_2PF_2IO_4$ POSS coatings, which cause an increase in fiber hydrophobicity, as well as a high drop in surface free energy, results in decreased adhesion of microorganisms and impaired conditions of their growth. In the modification process of cellulose fibers by DMeDHEU, the formation of covalent bonds between the finish and the cellulose macromolecules strengthens the less ordered amorphous regions, resulting in a decrease of fiber swelling. This inhibits the penetration of microorganisms into the fibers, where biodegradation takes place. By applying a combination of FAS and DMeDHEU finishes, a synergistic action of the components is obtained in the coating. In this case, the protective properties of the coating against biodegradation are much higher than those obtained by single component finishing.

Key words: cellulose, biodegradation, chemical modification, finishing, antimicrobial, water and oil repellent, durable press, soil burial test, microbial growth inhibition.

Chapter 1

INTRODUCTION

The study of biodegradation of cellulose fibers and their antimicrobial protection is a contemporary interdisciplinary research topic, which includes investigations in the fields of textile chemistry, polymer chemistry, biochemistry, microbiology, medicine, pharmacy, botany and ecology. Biodegradation is obviously an important and desirable process in waste textile materials, leading to their decomposition. On the other hand, it is strongly undesirable for textile products that are still in use, since an unpleasant odor, color stains and discoloration of fabrics occur during the biodegradation process, which significantly decreases the applicable value of textile products from hygienic as well as aesthetic points of view (Dring, 2003; Gao, 2008; Vigo, 1983). Moreover, in the case of textiles that are used in public places and transportation, there is a possibility of transmission of pathogenic microorganisms from one person to another and therefore a great chance of transmission of various infections and illnesses. Due to mould formation and rotting of textile fibers, biodegradation is also an urgent problem in conservation and restoration of textile products of great historical importance and in preservation of the cultural inheritance.

Cellulose fibers are classified as biodegradable. They are mainly comprised of cellulose, a high-molecular, water insoluble polymer of glucose. Since glucose is an important source of carbon, which is used by the microorganisms for their growth and development (Clarke, 1997; Glazer et al., 2001; Szostak-Kotowa, 2004), cellulose fibers are susceptible to bacteria and fungi as well as to algae, which, in the presence of moisture, heat and dirt, form a biofilm on the surface of the fibers, where they quickly multiply (Montegut et al., 1991). Due to the complexity of the nature of and interactions between different polysaccharides present in the cell wall of cellulose fibers and microorganisms, the biodegradation process is very complicated. In spite of the fact that the first scientific records of biodegradation of cellulose fibers date back to 1912 (Blackburn, 2005), this research field has not been fully investigated and, therefore, still remains open for further study.

The most active microorganisms that cause biodegradation of cellulose are generally found in water, soil and air. The most active fungi, therefore, belong to the genera *Chaetomium* sp., *Fusarium* sp., *Myrothecium* sp., *Memnoniella* sp., *Stachybotrys* sp., *Verticillium* sp., *Alternaria* sp., *Trichoderma* sp., *Penicillium* sp. and *Aspergillus* sp. The last two genera are of particular importance, since they include species that require relatively low humidity for their growth in comparison to other fungi. During biodegradation of cellulose, bacteria are of lesser importance than fungi, since they generally require conditions at which the fabric is saturated. Bacteria that degrade cellulose belong mostly to the genera *Cytohaga* sp., *Cellulomonas* sp., *Cellvibrio* sp., *Bacillus* sp., *Clostridium* sp. and *Sporocytophaga* sp. (Dring, 2003; Fischer et al., 1980; Itävaara et al., 1999; Raschle, 1989; Szostak-Kotowa, 2004). Bacteria degrade cellulose in different manner than fungi. While bacteria degrade cellulose fiber from its surface towards the inner parts of the fiber, degradation by fungi is just the opposite and proceeds from the inner parts of the fiber towards its outer side. In this case, the process of biodegradation starts at cracks in the surface or at the places where the fiber is cut off. Spores of fungi thus reach the lumen, from where their progression from the inner parts of the fiber toward the outer layers is possible (Szostak-Kotowa, 2004).

During microbiological attack of cellulose, fungi and bacteria produce cellulolytic enzymes, which catalyze the chemical reactions and therefore dramatically increase the rate and ratio of the cellulose biodegradation process. Enzymes are specialized proteins with different specificities and modes of action. Their activity is often referred to as "the lock and key" model (Figure 1), in which enzyme active sites that are of specific geometric shapes (key) fit exactly into the sites of substrate (lock). There are at least three distinct enzymes: endoglucanase, cellobiohydrolase and β-D-glukosidase (Clarke, 1997; Glazer, 2001; Szostak-Kotowa, 2004), which act synergistically to hydrolyze the cellulose to lower oligosaccharides, cellobiose and glucose. The first step of the cellulose biodegradation process is ascribed to the activity of endoglucanases, mostly carried out in the amorphous regions of cellulose fibers, where they randomly attack and hydrolyze the β-$(1\rightarrow4)$ bonds of cellulose to produce cello-oligosaccharides. Cellobiohydrolases degrade amorphous as well as crystalline cellulose, whereby they release disaccharide residues, i.e., cellobiose, from the non-reducing ends of cellulose molecules. The final step in the biodegradation of cellulose is the hydrolysis of cellobioses and soluble cello-oligosaccharides to glucose, which is caused by β-D-glukosidases (Clarke, 1997; Evans et al., 1998; Fischer et al., 1980; Itävaara et al., 1999). The mechanism of enzyme action in the cellulose biodegradation process is presented in Figure 2. Enzymatic degradation is reflected in a reduction of the degree of polymerisation of long cellulose chains, and damage to the fibers as well as a loss of breaking strength occur (Cao, 2002; Cao, 2004; Cao, 2005; Goynes, 1995; Seventekin, 1993; Szostak-Kotowa, 2004).

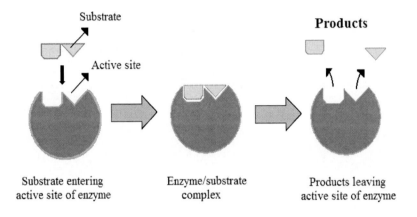

Figure 1. Mechanism of enzyme activity.

There are different possibilities of how to protect cellulose fibers against biodegradation. The chemical modification of fibers with antimicrobial agents is certainly one of the most effective methods. Of antimicrobial agents, biocides and biostats can be used, which differ in their chemical structure, the mechanism of antimicrobial action and their effectiveness (Dring, 2003; Gao, 2008; Vigo, 1983). Antimicrobial agents can be applied at different stages of fiber production. While in the spinning process of regenerative cellulose fibers, biocides are widely used as additives in the polymer master batch, in the case of natural cellulose fibers, chemical finishing with antimicrobial agents is the only possibility for fiber protection (Bajaj, 2002; Vigo, 1983). Regardless of the method of application, an antimicrobial agent kills microorganisms in contact with the fibers, resulting in a high inhibition of their growth.

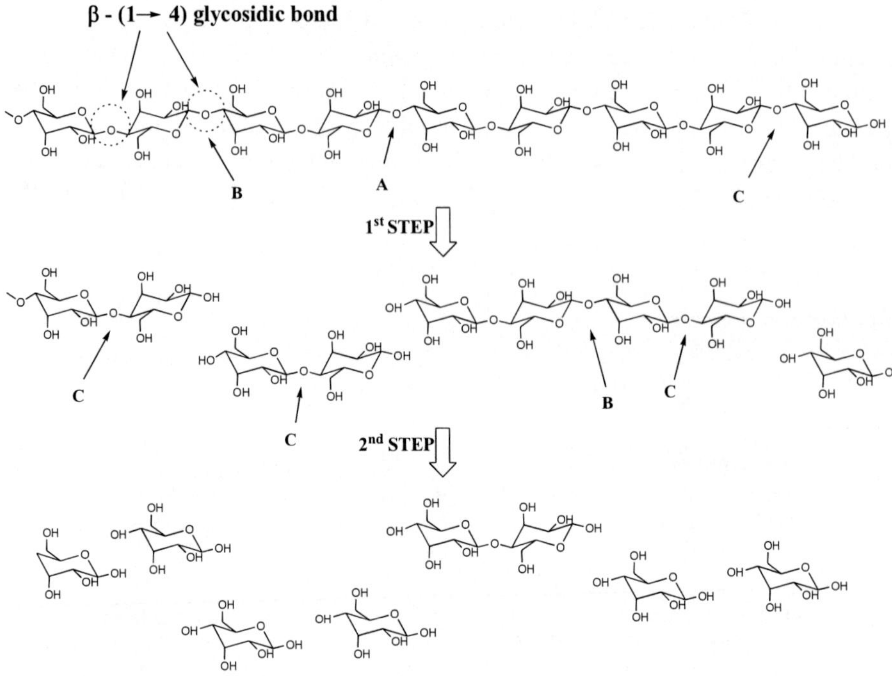

β - (1→ 4) glycosidic bond

Figure 2. Mechanism of enzymatic hydrolysis of cellulose; → denotes site of the attack of endoglucanase (A), cellobiohydrolase (B) and ß-glucosidase (C).

On the other hand, the rate and degree of cellulose biodegradation can also be successfully inhibited by application of chemical finishes, which are not active as antimicrobial agents, but their presence on the fibers significantly impairs the conditions for microbial growth. It could be denoted as passive antimicrobial activity. In this fiber modification process, it is taken into consideration that the amorphous regions of cellulose are much more susceptible to attack by microorganisms than crystalline cellulose, since the degradation process occurs at sites where fibers are in direct contact with the microorganisms, and that the activity of microorganisms is largely influenced by the environmental conditions, in which the presence of sufficient moisture is of great importance (Buschle-Diller et al., 1994; El-Morsy, 2001; Montegut, 1991; Park, 2004). It can thus be expected that any kind of chemical modification of cellulose fibers that leads to a strengthening of their amorphous regions and an increase in fiber hydrophobicity will effectively decrease fiber biodegradation. The first condition can be obtained by applying easy care and durable press finishes and the second with the use of water and oil repellent finishes. It is very important to note that both types of finish can modify the fiber chemical properties to such an extent that it causes reduced compatibility between enzyme active sites and the sites of cellulose substrate. Moreover, water and oil repellent finishes form a low energy polymer network on the fiber surface, which greatly decreases the adhesion between microorganisms and cellulose macromolecules (Zhao, 2004), as well as enzyme-cellulose interactions (Figure 3). Since the presence of water and oil repellent finishes greatly increase the hydrophobicity and oleophobicity of fibers in comparison to the untreated ones, this also greatly decreases the moisture in the fibers, which is crucial for the growth of microorganisms.

Figure 3. A drop of water, including cellulolytic enzymes, on the surface of cotton fabric coated with water and oil repellent finish.

This work consisted of two parts. In the first, we focused on the biodegradation process of cotton fabric, using a soil burial test to determine the resistance of cellulose fibers to the action of soil microflora. The rate and degree of the biodegradation process were demonstrated by scanning electron microscopy, measurement of the degree of polymerization and the fiber breaking strength, as well as by Fourier transform infrared spectroscopy. The second part of the work was devoted to an assessment of the active and passive antimicrobial activity of different finishes on cellulose fibers. A silver-based finish, including AgCl salt, embedded in the organic-inorganic matrix was used for preparation of the active antimicrobial properties of cellulose fibers. For passive antimicrobial activity we used a non-formaldehyde containing product based on imidazolidinone, a fluoroalkylfunctional water-born siloxane (FAS), 1H, 1H, 2H, 2H-perfluorooctyltriethoxysilane (PFOTES) in combination with di-ureapropyltriethoxysilane [bis(aminopropyl)terminated-polydimethylsiloxane (1000) (PDMSU), as well as aminopropyl-perfluoroisooctyl polyhedral oligomeric silsesquioxane ($AP_2PF_2IO_4$ POSS) in combination with a diisocyanatohexyl (DICH) cross-linker. The presence of finishes on the cellulose fibers was demonstrated by scanning electron microscopy and energy-dispersive X-ray spectroscopy, atomic force microscopy, inductively coupled plasma mass spectroscopy, Fourier transform infrared spectroscopy, X-ray photoelectron spectroscopy and liquid contact angle measurements. The antimicrobial activity of finishes was investigated with the use of a soil burial test, as well as appropriate microbiological tests in which fungicidal activity was estimated for the fungi *Aspergillus niger* and *Chaetomium globosum* according to the modified DIN 53931 Standard Method, and bactericidal activity was estimated for the Gram-negative bacterium *Escherichia coli* using three standard methods: ISO 20645:2004 (E), AATCC 100-1999 and the EN ISO 20743:2007 Transfer Method. By employing the EN ISO 20743:2007 Transfer Method, it was possible to confirm the beneficial effect of low surface energy FAS, PFOTES and $AP_2PF_2IO_4$ POSS finishes on the reduction of bacterial growth on treated cotton fabrics.

Chapter 2

MATERIALS FOR FINISHING AND APPLICATION METHODS

Plain-weave 100 % cotton woven fabric with a mass of 164 g/m^2 was used in the experiments. In a pre-treatment process the fabric was bleached in an H$_2$O$_2$ bath, mercerised in a NaOH solution and neutralized in a diluted CH$_3$COOH solution.

As an antimicrobial agent, iSys AG was used in combination with iSys MTX (CHT, Germany). The former is a dispersion agent containing AgCl (AG) and the latter is a reactive organic-inorganic binder (RB).

As water and oil repellent agents, we used Dynasylan F 8815 (Degussa, Germany), which is a fluoroalkylfunctional water-born siloxane (FAS), 1H, 1H, 2H, 2H-perfluorooctyltriethoxysilane (PFOTES) (ABCR), in combination with di-ureapropyltriethoxysilane (bis(aminopropyl)terminated-polydimethylsiloxane (1000) (PDMSU), as well as aminopropyl-perfluoroisooctyl polyhedral oligomeric silsesquioxane (AP$_2$PF$_2$IO$_4$ POSS), in combination with diisocyanatohexyl (DICH) cross-linker. Synthesis of PDMSU and AP$_2$PF$_2$IO$_4$ POSS was reported previously (Jerman et al., 2008a, b; Lavrenčič-Štanger et al., 2005; Vince et al., 2006).

As an easy-care and durable-press finish, a non-formaldehyde containing product based on imidazolidinone (CHT, Tübingen), 1,3-dimethyl-4,5-dihydroxyethylene urea (DMeDHEU) was used. All products can be mixed with water to any desired concentration.

The AG-RB finish was applied to the cotton fabric by the exhaustion method. The AG-RB sol-gel solution was therefore prepared using 3.0 g/l AgCl (which is equivalent to 0.15 % o.w.f.) in combination with 15 g/l RB with a liquid ratio of 1:50. The sample was immersed in the solution and left at room temperature with occasional stirring until equilibrium was achieved. The sample was then wrung to a wet pick-up of 80 ± 1 %, dried at 120 °C and cured at 150 °C for 1 minute.

The cotton fabric was finished by 10 % FAS and 80 g/l DMeDHEU, as well as by their mixture using the pad-dry-cure method, including full immersion at 20 °C, wet pick-up of 80 ± 1 % at 20 °C, drying at 100 °C and curing at 150 °C for 5 minutes. The cross-linking reaction of the DMeDHEU finish with the cellulose hydroxyl groups was catalyzed with 15 g/l hydrated magnesium chloride (Aldrich).

The application of the 4 % PFOTES-PDMSU and the 4 % $AP_2PF_2IO_4$ POSS-DICH sols to the cotton fabric was carried out by the pad-dry-cure method, including full immersion, wet pick-up of 100 ± 2 % at 20 °C, drying at 100 °C and curing at 140 ° for 15 min.

Using the dip-coating technique, FAS, PFOTES-PDMSU and $AP_2PF_2IO_4$ POSS-DICH sols were applied to smooth and polished aluminum AA 2024 alloy substrates for contact angle measurements. Heat treatment was performed at 140 °C for 15 min. The precursor concentrations in sols were the same as those used for the application to the cotton fabric.

Chapter 3

ANALYTICAL METHODS FOR THE
EVALUATION OF MATERIALS AND FINISHES

3.1. SOIL BURIAL TEST

Determination of the resistance of finished and unfinished cotton fabrics to the action of soil microflora was carried out by the soil burial test, according to ISO 11721-1:2001 and ISO 11721:2003 standards. In this standard process, a container was filled with commercial grade compost. The water content of the soil was 60 ± 5 % of its maximum moisture retention capacity. It was held constant during the experiment by spraying with water. The pH of the soil was between 4.0 and 7.5. Cotton fabric samples were buried in the soil for periods of 3, 6, 9 and 12 days. After the defined incubation time, the samples were removed from the test soil, lightly rinsed with running tap water and immersed in 70 % ethanol for 30 minutes before air drying.

3.2. SCANNING ELECTRON MICROSCOPY (SEM) AND ENERGY-DISPERSIVE X-RAY SPECTROSCOPY (EDXS)

The morphology and composition of the coatings on the cotton fabrics were investigated in a JEOL JSM 5800 scanning electron microscope (SEM) equipped with an Oxford-Link ISIS 300 EDXS system with an ultra-thin window Si(Li) detector. The samples for SEM and EDXS analysis were coated with a $\approx$ 30-nm-thick carbon layer to ensure sufficient electrical conductivity and to avoid charging effects. Analyses were performed using a 10-keV electron beam, 200 to

500 pA beam current and X-ray spectra acquisition under a 35 $^{\circ}$ take-off angle. SEM micrographs were recorded using both secondary electron (SE) and backscattered electron (BSE) imaging modes. BSE compositional (Z-contrast) imaging was applied to emphasize and expose the difference between the added particles and the cotton fiber-matrix.

3.3. ATOMIC FORCE MICROSCOPY (AFM)

The topography of untreated and coated cotton fibers was measured by atomic force microscopy (AFM) with the use of atomic force microscope Solver Pro (NT-MDT Co.). Scanning in oscillating mode was performed. All images were obtained at ambient condition. The roughness parameter (rms) was calculated from the AFM images taken over 5 x 5 mm. Before the roughness calculation, a second-order plane was subtracted from each AFM image to take into account the curvature of the fibers. Two places were analyzed on each sample.

3.4. BREAKING STRENGTH

Breaking strength was measured with an Instron 5567 dynamometer in accordance with SIST ISO 5081:1996. The relative reduction in breaking strength, q_{red}, of the buried cotton samples compared with the unburied ones was calculated from the mean value of the breaking strength of ten specimens, using the following relationship:

$$q_{red,t} = \frac{F_t}{F_{t0}} \qquad (1)$$

where $q_{red,t}$ is the loss of breaking strength of the buried cotton sample after burial time t, F_t is the breaking strength of the buried cotton sample after burial time t, and F_{t0} is the breaking strength of the unburied cotton sample. Before testing, the samples were conditioned at 65±2 % relative humidity and 20±1 $^{\circ}$C for 24 hours.

3.5. DEGREE OF POLYMERIZATION (DP)

The DP of the cellulose samples dissolved in Cuoxam, a solution of cupric hydroxide in aqueous ammonia $[Cu(NH_3)_4](OH)_2$, was determined viscosimetrically using an Oswald shear dilution viscometer.

3.6 FOURIER TRANSFORM INFRARED (FT-IR) SPECTROSCOPY

FT-IR spectra were obtained on a Brucker IFS 66/S spectrophotometer equipped with an attenuated total reflection (ATR) cell (SpectraTech) with a Ge crystal ($n = 4.0$). The spectra were recorded over the range $4000 - 600$ cm^{-1}, with a resolution of 4 cm^{-1} and averaged over 128 spectra. Before measurement, the studied samples were dried for 5 hours at 100 ^{o}C.

3.7. X-RAY PHOTOELECTRON SPECTROSCOPY (XPS)

X-ray photoelectron spectroscopy (XPS or ESCA) analyses were carried out on a PHI-TFA XPS spectrometer (Physical Electronics Inc). The analyzed area was 0.4 mm in diameter and about 3-5 nm in depth. This high surface sensitivity is a general characteristic of the XPS method. Sample surfaces were excited by X-ray radiation from a monochromatic Al source at a photon energy of 1486.6 eV. C 1s, F 1s, O 1s, N 1s and Si 2p spectra were acquired with an energy resolution of about 1.0 eV with an analyzer pass energy of 58 eV. Quantification of surface composition was performed from XPS peak intensities measured on three different spots of the sample, taking into account the relative sensitivity factors provided by the instrument manufacturer (Moulder, 1995).

3.8. INDUCTIVELY COUPLED PLASMA MASS SPECTROSCOPY (ICP-MS)

The concentration of Ag in the finished cotton bulk samples was determined by ICP-MS on a Perkin Elmer SCIED Elan DRC spectrophotometer. A sample of 0.5 g was prepared in a Milestone microwave system by acid decomposition using

65 % HNO_3 and 30 % H_2O_2. Three measurements were made for each sample, and the Ag concentration was given as a mean value.

3.9. MICROBIOLOGICAL TESTS

3.9.1. Fungicidal Activity

The fungicidal activity of the Ag-RB-treated cotton samples was estimated for the fungi *Aspergillus niger* (ATCC 6275) and *Chaetomium globosum* (ATCC 6205) according to the modified DIN 53931 Standard Method, in which synthetic nutrient-poor agar (SNA) (Nirenberg 1976) consisting of 1 g of KH_2PO_4, 1 g of KNO_3, 0.5 g of $MgSO_4 \times 7 H_2O$, 0.5 g of KCl, 0.2 g of glucose, 0.2 g of saccharose and 15 g of technical agar per 1 l distilled water was used instead of the prescribed malt-extract agar (MEA). SNA is a less nutritious cultural medium, allowing a more delicate colony growth and easier evaluation of the antifungal activity of Ag. Thirty µl of spore-suspension (10^5/ml) were spread on each SNA plate. The inoculated plates were incubated at 29 °C for 24 hours. Samples of cotton fibers 5 × 5 cm were then placed on the medium and incubated at 29 °C for 7 and 14 days. After incubation, fungicidal activity was determined in terms of mycelial growth on and below the surface of the cotton fibers and the intensity of sporulation. To determine fungal development below the finished samples, the cotton samples were removed from the agar plate and the medium was examined microscopically. A droplet of cotton blue dissolved in lactic acid was placed on the medium and covered with a glass slip, allowing the detection of the stained fungal mycelium at low magnification. The degree of fungal growth was ordered in 8 grades from 00 to 5, where 00 indicates no growth, 0 fungal growth outside an inhibition zone surrounding the cotton specimen, [0] fungal growth up to the specimen's edge, 1 fungal growth only on and below the specimen's edge, 2 fungal growth on and below less than 25 % of the specimen, 3 fungal growth on and below 25 – 75 % of the specimen, 4 fungal growth on and below more than 75 % of the specimen and 5 100 % overgrowth of the specimen. The intensity of sporulation was assessed using the following symbols: – means clear, without mycelium; + weak, only mycelium; ++ noticeable growth, partly spores; and +++ strong overgrowth, extensive spore formation.

3.9.2. Bactericidal Activity

The antibacterial activity of coated cotton samples was estimated for the Gram-negative bacterium *Escherichia coli* (ATCC 25922) using three different standard methods: ISO 20645:2004 (E), AATCC 100-1999 and the EN ISO 20743:2007 Transfer Method.

For the ISO 20645:2004 (E) standard method, known as the Agar diffusion plate test, two-layered agar plates were prepared. The lower agar layer consisted of 10 ml of ordinary agar; the upper layer consisted of 5 ± 1 ml agar inoculated with bacteria, whereby one ml of bacteria working solution with a concentration of $1\text{-}5 \times 10^8$ CFU was added per 150 ml of agar. Circular pieces of cotton, 25 ± 5 mm in diameter, were uniformly pressed on the agar and incubated for 24 hours at 37 ± 1 °C. After incubation, assessment was based on the absence or presence of bacterial growth in the contact zone between the agar and the sample and on the possible appearance of an inhibition zone, which was calculated from:

$$H = \frac{D-d}{2} \tag{2}$$

where H is the inhibition zone in mm, D is the total diameter of the cotton specimen and inhibition zone in mm, and d is the diameter of the specimen in mm. For bacterial growth, the contact zone under the samples was also determined with a microscope at 20-times magnification. For the standard method, the inhibition zone was measured in mm and the degree of bacterial growth was estimated in the nutrient medium under the specimen. The antibacterial effect of the studied samples was described as "good", "limited" or "insufficient". All tests were performed in duplicate.

According to the AATCC 100-1999 standard method, circular swatches of finished cotton samples, 4.8 cm in diameter, were put into a 250 ml Erlenmeyer flask and inoculated with 1.0 ml of a nutrient broth culture containing $1\text{-}2 \times 10^5$ CFU of bacteria. An unfinished cotton sample was used as a control. After incubation at 37 °C for 24 hours, the bacteria were eluted from the swatches by shaking them in 100 ml of neutralizing solution for 1 minute. After making serial dilutions with sterilized water, the suspensions were plated on nutrient agar and incubated at 37 °C for 24 hours. The number of bacteria forming units (CFU) was then counted, and the reduction of bacteria, R, was calculated from:

$$R = \frac{(B-A)}{B} 100 \quad (\%) \tag{3}$$

where A is the CFU recovered from the inoculated cotton sample swatch in the jar incubated over the desired contact period (24 hours), and B is the CFU recovered from the inoculated cotton sample swatch in the jar immediately after inoculation (at "0" contact time). Three treatments were performed on two samples for each finished cotton fabric.

In addition, the reduction of bacterial growth on the finished samples was also estimated using the EN ISO 20743:2007 Transfer Method. This method enables an assessment of the bacterial reduction that is caused not only by the presence of antibacterial active agents (i.e., nanosilver particles) in the finishes, but could stem from the low surface energy of the oleophobic finished cotton (caused by FAS), which prevents or at least hinders the adhesion of bacteria and their consequent growth and the formation of a biofilm on the finished fabrics. The latter effect is called 'passive antimicrobial activity' to distinguish it from the inherent antibacterial effect of various chemical agents embedded in finishes, such as the silver nonoparticles used in this study. According to the EN ISO 20743:2007 Transfer Method, the agar plates were inoculated with 1 ml of a nutrient broth culture containing $1\text{-}3 \times 10^6$ colony forming units of Gram-negative bacteria *Escherichia coli* (ATCC 25922). The swatch of the test sample (3.8 cm in diameter) was plated on the agar surface and pressed down with a 200 g cylindrical weight for 60 ± 5 s. The test sample was then removed from the agar surface, placed in a 100 ml container with the transferred surface face up and incubated at 37 °C for 24 h in a humidity chamber. After incubation, 20 ml of neutralizing solution was poured on the test sample and it was shaken vigorously for 1 min. After making serial dilutions with sterilized water, the suspensions were plated on nutrient agar and incubated at 37 °C for 24 hours. The reduction of bacteria growth, R_A, on the finished sample in comparison to the unfinished one was calculated as follows:

$$R_A = \frac{(U-T)}{U} 100 \quad (\%) \tag{4}$$

where U is the number of bacteria forming units on the untreated sample after 24 hours incubation and T the number of bacteria forming units on the treated samples under the same conditions. Three treatments were performed on two samples for each fabric sample.

3.10. ASSESSMENT OF WWETTING PROPERTIES

The static (equilibrium) contact angles of water (W), formamide (FA) and diiodomethane (DIM) on the coatings prepared on Si wafers and finished cotton samples were made on a DSA 100 contact angle goniometer (Krüss, Germany), which works on the principle of the goniometer-sessile drop technique. From the image of the deposited liquid drop on the surface, the drop contour was analyzed and the contact angle was determined by using the Young-Laplace fitting, which is the theoretically most exact method. With this method, the complete drop contour was evaluated taking into account the drop deformation caused by liquid weight, which, in addition to interfacial effects, also contributed to the drop shape. Liquid drops of 8 μl were deposited on different spots of the coated substrates to avoid the influence of roughness and gravity on drop shape. The average values of contact angles in 30 – 60 seconds from the deposition of the drop were calculated from at least five measurements on glass plates and from at least ten measurements on the studied fabric, minimizing errors due to roughness and heterogeneity. Contact angle measurements were carried out at 20 °C and ambient humidity.

Van Oss and co-workers approach (van Oss et al., 1988a, b) was used for determination of the total surface free energy of the coatings, resolved to the corresponding apolar Lifshitz-van der Waals component, γ_S^{LW}, which includes London dispersion, induction (Debye) and orientation (Keeson) interactions, and a polar component, γ_S^{AB}, due to electron-donor, γ_S^-, and electron-acceptor, γ_S^+, interactions. According to this theory, the solid surface free energy components γ_S^{LW}, γ_S^- and γ_S^+ can be obtained from the following expression:

$$W_A = (1 + \cos\theta)\gamma_L = 2\sqrt{\gamma_S^{LW}\gamma_L^{LW}} + 2\sqrt{\gamma_S^+\gamma_L^-} + 2\sqrt{\gamma_S^-\gamma_L^+} \tag{5}$$

if contact angles of three different liquids with known γ_L^{LW}, γ_L^- and γ_L^+ are measured on the same solid surface. Practice suggests that when using Equation (4), one of the chosen liquids should be non-polar. To perform the van Oss calculations, DIM (γ_L^{LW} = 21.6 mJ/m^2, γ_L^- = 0,0 mJ/m^2, γ_L^+ = 0.0 mJ/m^2), FA (γ_L^{LW} = 39.0 mJ/m^2, γ_L^- = 39.6 mJ/m^2, γ_L^+ = 2.28 mJ/m^2) and W (γ_L^{LW} = 21.8 mJ/m^2, γ_L^- = 25.5 mJ/m^2, γ_L^+ = 25.5 mJ/m^2) were used.

3.11. ASSESSMENT OF WASHING FASTNESS

The washing fastness of coatings was determined by repetitive washing in an AATCC Atlas Launder-O-Meter Standard Instrument, which is widely used for evaluating laundry results on a laboratory scale. One wash in a Launder-O-Meter (ISO 105-C01:1989(E) standard method) provides an accelerated washing treatment corresponding to five home washings. The finished fabric samples were washed repetitively up to 10 times; the duration of the washing cycles was 30 min and was carried out in a solution of SDC standard detergent of concentration 5 g/l, previously heated to 40 °C, to give a liquid ratio of 50:1. After washing, the samples were rinsed in cold distilled water and then held under cold tap water for 10 min, squeezed and dried at room temperature. After drying, samples were also heat-treated by ironing at 190 °C for 10 s. The quality of the coatings was assessed after the first and tenth washing cycles.

Chapter 4

BIODEGRADATION OF UNTREATED CELLULOSE FIBERS

4.1. FIBERS BURIED IN SOIL

The biodegradation process of cellulose fibers was carried out by the soil burial test and the results are presented in Figure 4. Photos of samples removed from the test soil after different incubation times showed that the speed of the rotting process caused by microorganisms in the soil accelerated over the time of burial. After 12 days of burial, the cotton fabric was degraded to such an extent that it fell to pieces. SEM images thus revealed major morphological changes due to the decomposition action of the soil microflora. In the initial stage of the experiment (0 day), the surface of the cotton fabrics was very smooth. After 6 days of soil burial, the development of superficial cracks in the case of the fibers could be observed. The intensity of this morphological damage increased with increasing soil burial time, whereby serious disintegration and defibrillation of the untreated fibers could be observed after 12 days of soil burial to such an extent that individual macrofibrils could be seen.

The biodegradation process was additionally demonstrated by breaking strength and DP measurements. The results shown in Figure 5 reveal that the breaking strength of the cotton samples rapidly decreased with increasing time of burial and reached a value of 0.002 after 9 days of incubation. The results also show that the loss of breaking strength was directly related to rupture of the β-$(1\rightarrow4)$ glycosidic bonds of the cellulose macromolecules (see Figure 2), resulting in a marked decrease of DP from 1923 for an unburied sample to 1551 for a sample buried for 12 days.

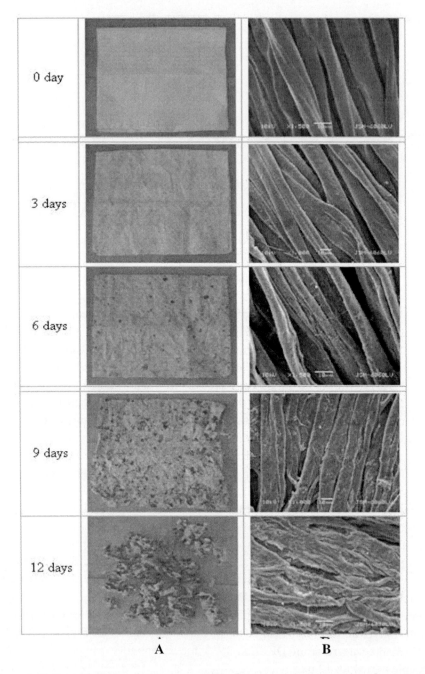

Figure 4. Photographs (**A**) and SEM images (**B**) of untreated cotton samples after 0, 3, 6, 9 and 12 days of the burial test.

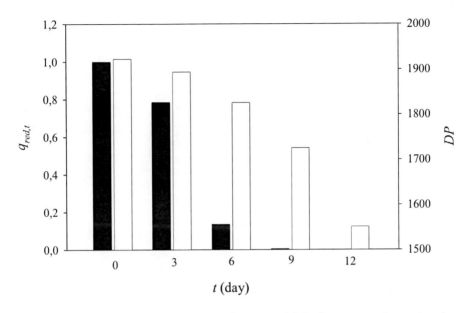

Figure 5. Plots of the loss of breaking strength, $q_{red,t}$, and DP of cotton samples against the burial time, t. ■ - $q_{red,t}$, □ - DP.

4.2. FT-IR SPECTRA OF DEGRADED FIBERS

Important information about the biodegradability of cellulose fibers can be provided by infrared spectral analysis (Figure 6). An inspection of the measured ATR spectra of cotton samples after 12 days of burial revealed that microbial biodegradation led to cellulose structural changes. To provide a clear view of the changes appearing during the biodegradation process, the difference ATR spectra, obtained by subtracting the spectrum of unburied cellulose from that of the the buried one, were examined. In Figure 3A, negative band absorptions occur at 3340 cm⁻¹ due to OH stretching, at 2918 and 2853 cm⁻¹ due to CH and CH_2 stretching and at 1456, 1429, 1370, 1336, 1318, 1280, 1160, 1105, 1053 and 1029 cm⁻¹ due to C-C and C-O stretching vibrations, skeletal vibrations and ring vibrations in the cellulose fingerprint region as a result of the decreased intensities of these bands after biodegradation. Hulleman et al. (1994) suggested that the decrease of band intensities in the region 1500-900 cm⁻¹ may be attributed to a decrease in cellulose crystallinity. This finding is in agreement with that obtained by Nelson et al. (1964), whereby the intensity of bands at 1429, 1370, 1335, and 1315 cm^{-1} (which were assumed to be very sensitive to changes in cellulose crystallinity and lattice type) decreased with decreasing crystallinity. This additionally confirmed the characteristic decrease in the 900 cm^{-1} band and its broadening after 12 days of burial (Colthup et al., 1990; Nelson et al., 1964).

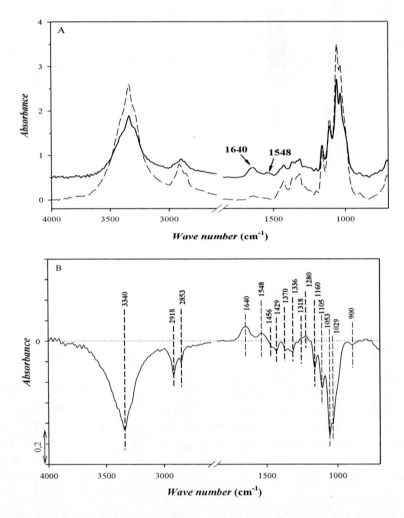

Figure 6. Measured infrared ATR spectra (A) of cotton samples before (—) and after (---) 12 days of burial. Difference ATR spectrum (B) obtained by subtracting the spectrum of unburied cellulose from that of the buried one.

Simultaneously, significantly increased broad bands appear at 1640 and 1548 cm^{-1}. This indicates that very important structural changes appear in the spectral region from 1700-1500 cm^{-1}, reflected in the change of intensity of bands at 1640 and 1548 cm^{-1}. A detailed analysis of the spectra showed that the significant increase of the band at 1640 cm^{-1}, obtained for cotton samples buried for 12 days, was not caused by an increase of adsorbed water onto the degraded cellulose (Tomšič et al., 2007). Further inspection of the bands at 1640 and 1548 cm^{-1} also shows that their position and shape are not representative of the spectral absorption of aldehyde or carboxylic functional groups, which are produced in the cellulose biodegradation process. They seem to be more related to Amides I and II (Socrates, 2001; Vince et al., 2006). Their appearance in the cellulose structure could be explained by the presence of secondary polyamides due to the proteins that are produced during the growth of microorganisms on the fibers, which can be adsorbed on the degradable cellulose macromolecules. Microbial growth thus results in an increase of the bands at 1640 and 1548 cm^{-1}. These results are in agreement with the vibrational spectra of bacteria grown on a culture media (Socrates, 2001).

Chapter 5

INHIBITION OF CELLULOSE BIODEGRADATION BY CHEMICAL MODIFICATION

5.1. ACTIVE AND PASSIVE ANTIMICROBIAL PROTECTION

The influence of chemical modification of cellulose fibers with AG-RB, FAS, DMeDHEU and FAS-DMeDHEU coatings on the biodegradability of cellulose fibers is presented in Figure 7. It is clearly seen from the results that the chemical modification of cellulose fibers with all the used chemical finishes inhibits the rate and degree of cellulose biodegradation, resulting in lower morphological damage after 12 days of soil burial in comparison with the untreated sample. These results are in accordance with the results of the loss of breaking strength that is directly influenced by the degree of sample degradation. It can be seen from Figure 8 that the AG-RB coating of the cellulose caused the highest inhibition of fiber biodegradation, followed by FAS-DMeDHEU, DMeDHEU and FAS coatings. Excellent antimicrobial properties of the AG-BR coating were expected, since it included AgCl, a well-known biocide, which provided active antimicrobial activity of modified fibers. However, the very high inhibition of the biodegradation process of cellulose fibers modified by DMeDHEU and FAS was not expected, since neither of the finishes used are antimicrobial agents. Nevertheless, the results indicated that the presence of either of the finishes on the cellulose fibers strongly impaired the conditions for the growth of microorganisms. This property was designated passive antimicrobial activity. Active and passive antimicrobial activity was proved by the appropriate microbiological tests. The properties of modified cellulose fibers were determined on the basis of FT-IR and EDXS spectroscopy, as well as sample wettability using static contact angle measurements.

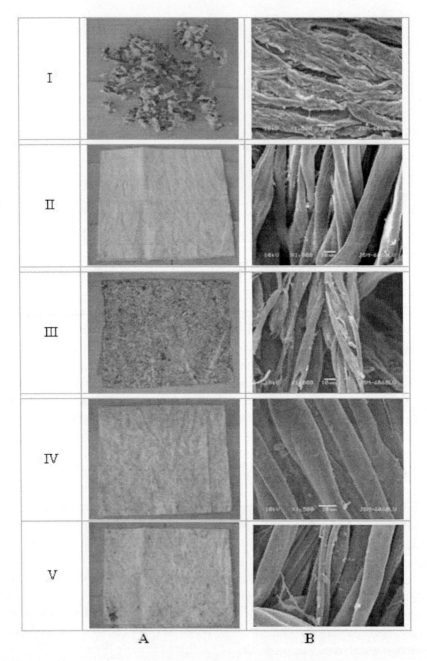

Figure 7. Photographs (**A**) and SEM images (**B**) of untreated cotton sample (I) and samples coated with AG-RB (II), FAS (III), DMeDHEU (IV) and combination of FAS-DMeDHEU (V) finishes after 12 days of burial test.

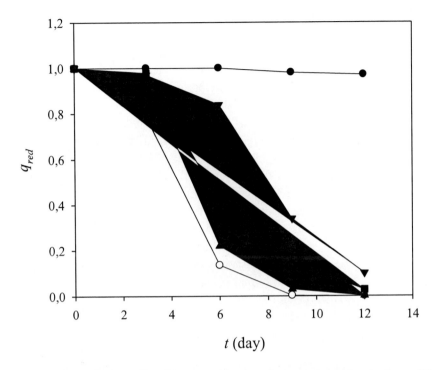

Figure 8. Plot of the loss of breaking strength, q_{red}, of untreated and chemically modified cotton samples against burial time, t. O - untreated sample, ● - coated with AG-RB, ▲ - coated with FAS, ■ - coated with DMeDHEU, ▼ - coated with combination of FAS-DMeDHEU.

5.1.1. Active Antimicrobial Activity of AG-RB Finish

Mechanism of Biocidal Activity of AgCl in the Oxide Matrix

An AG-RB coating with the added AgCl salt was used for imparting antimicrobial properties to the cellulose fibers. Silver-based finishes are an important group of anti-microbial agents (Kissa, 1984; Simončič, 2003; Schindler & Hauser, 2004; Gao & Cranston, 2008), which act as an effective bactericide and fungicide even at low concentrations. These compounds are not chemically bonded onto the textile fibers and their antimicrobial activity is attributed to their gradual and persistent release from the textile into surroundings, where they act as a poison to a wide spectrum of bacteria and fungi. The anti-microbial mechanism of AgCl is poorly known. However, it seems that Ag^+ ions released from AgCl act toxically towards unicellular micro-organisms by their binding to negatively-charged carboxyl, thiol and phosphate side groups in microbial cells (Gupta et al., 1998; Matsumura et al., 2003; Kim et el., 2007; Lee et al., 2007) and, among other things, also by interacting with thiol groups of proteins. Due to the abundance of sulphur-containing proteins on the bacterial cell membrane, Ag^+ cations can react with sulphur-containing proteins inside or outside the cell membrane, which in turn affects bacterial cell viability. It was also proposed that the released Ag^+ cations can interact with phosphorus moieties in DNA, resulting in deactivation of DNA replication, as well as with sulphur-containing proteins, leading to the inhibition of enzyme functions. At high concentrations, silver atoms and silver ions are also toxic to fungi (Jung et al., 2007). It is also believed that silver is biocompatible and non-toxic to human cells at concentrations effective against microorganisms (Kusnetsov et al., 2001; Lee & Jeong, 2005), when in the form of non-agglomerated and well dispersed particles.

To obtain uniform dispersion, embedment and controlled release of Ag^+ particles, RB was used in combination with AG. RB is a reactive organic-anorganic binder forming an oxide matrix on the fibers in which silver particles can be physically bonded (Mahltig, 2005). The oxide matrix is biologically inert, and it does not represent a food source for micro-organisms. Previous studies have shown that the presence of an oxide matrix increases the concentration of bounded silver as well as its uniform distribution and prolongs the time of Ag^+ release into the environment (Akkopru & Durucan, 2007; Mahltig et al., 2004; Tomšič et al., 2008; Wang et al., 2006; Xing et al., 2007).

Structure of Cellulose Fibers Modified by AG-RB Finish

Chemical modification of cellulose by the AG-RB coating was demonstrated by SEM micrographs (Figure 9) as well as EDXS and ATR spectra (Figure 10). Morphological changes and the distribution of the silver particles on the AG-RB finished cotton fabric can be observed from the SEM micrographs of the cotton fabric before (Figure 9A) and after (Figure 9B) the application of the AG-RB finish. The silver particles were spherical in shape and from 100 to 500 nm in size. Their presence on the finished cotton fibers was confirmed by the EDXS spectrum (Figure 10A). The bulk concentration of Ag particles on the cotton fibers, which reached 130 mg/kg was determined on the basis of ICP-MS analysis (Tomšič et al., 2009). The ATR spectrum of the AG-RB coating (Figure 10B) revealed bands of Si-O-Si linkages at 1130, 1075 and 1025 cm^{-1} (shoulder), showing that a silica network was formed during the condensation process, capable of incorporating Ag particles. It should be noted that the AG-RB coating was studied on a Si wafer since Si-O-Si bands at

1075 and 1025 cm^{-1} coincided with the strong bands of the cellulose fingerprint (Vince et al., 2006; Fir et al., 2007; Tomšič et al., 2008, 2009).

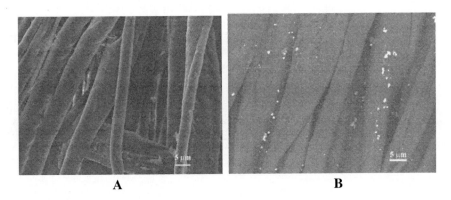

Figure 9. SEM images of untreated cotton fibers (**A**) and fibers coated with AG-RB finish (**B**).

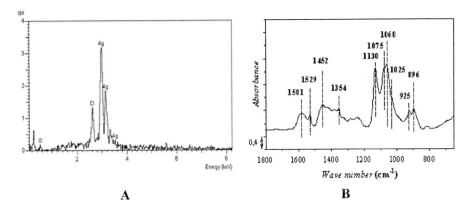

Figure 10. Measured EDXS spectrum (**A**) of cotton fibers coated with AG-RB finish, and ATR spectrum of AG-RB coating (**B**) on a Si wafer in the 1800-650 cm^{-1} spectral region.

Antibacterial and Antifungal Activity of AG-RB Coating

The results of the antibacterial and antifungal activity of AG-RB coating on the cellulose fibers are presented in Figure 11 and in Table 1. A comparison of photos of the growth of *E. coli* on nutrient medium covered with an untreated cotton sample (Figure 11, IA) and a cotton sample coated with AG-RB finish (Figure 11, IB) revealed that the presence of AgCl in the coating strongly suppressed bacterial growth. The release of Ag^+ particles from the coating into its surrounding resulted in the formation of an inhibition zone greater than 1 mm. It is clearly seen from the microscopic images of the *E. coli* growth on the nutrient medium that no bacterial colonies (small shadowed spots) were present in the inhibition zone near the edge of the removed sample (Figure 11, IC), nor in the nutrient medium under the sample itself. Contrary to the coated cotton sample, an insufficient antibacterial effect was expectedly obtained for the untreated cotton sample (Figure 11, IA), caused by a lack of an inhibition zone and a heavy bacterial growth on the medium under the sample. The excellent antimicrobial efficiency of the AG-RB coating was demonstrated by the reduction of bacteria, *R*, shown by the AATCC 100-1999 standard method, whereby 100 % of bacterial reduction was obtained for the coated cotton sample and no bacterial reduction for the untreated cotton sample.

It can also be seen from Figure 11 (II and III) and Table 1 that the concentration of AgCl in the coating was high enough to obtain antifungal properties of the modified cellulose fibers. Both fungi *C. globosum.* and *A. niger* grew all over the untreated cotton sample (Figure 8, IIA and IIIA), but their growth was strongly inhibited in the case of the coated cotton sample (Figure 11, IIB and IIIB). The microscopic observation of submersed mycellal growth on the nutrient medium near the edge of the removed cotton sample revealed that the antifungal activity of AG-RB coating was lower for *A. niger* than for *C. globosum*. While the sample was totally unaffected by *C. globosum* (Figure 11, IIC), an inspection of the sample in contact with the medium previously inoculated by *A. niger* revealed that, in spite of the fact that there was no fungal growth present on the sample, a restricted mycelium with chains of candidia was found on some spots of the medium under the sample (Figure 11, IIIC). Nevertheless, these results clearly indicated that fungal growth was strongly attacked by the modified cotton fibers but not completely inhibited.

The results of the microbiological tests are in accordance with the results of the biodegradation process of cotton fabric coated with the AG-RB finish. The very low damage caused by microorganisms in the soil even after 12 days of burial is certainly influenced by the active and highly effective antimicrobial and antifungal activity of AgCl in the coating.

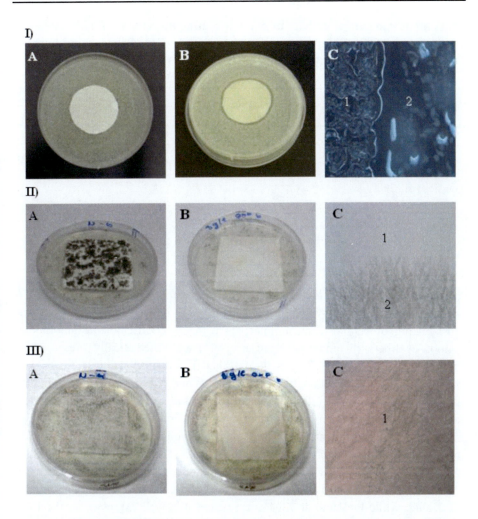

Figure 11. Growth of bacterium *E. coli* (I) and fungi *C. globosum* (II) and *A. niger* (III) on the nutrient medium covered with untreated cotton sample (A) and cotton sample coated with AG-RB finish (B). C) Microscopic observation of bacterial growth (IC) and of submersed mycelial growth of *C. globosum* (IIC) and *A. niger* (IIIC) on the nutrient medium near the edge of the removed cotton sample (1 – under the sample, 2 – in the surrounding of the sample). The antibacterial test was carried out according to the ISO 20645:2004 (E) standard method, and the antifungal test was carried out according to the modified DIN 53931 standard method.

Table 1. Antifungal activity of AG-RB coating on cellulose fibers for *C. globosum* and *A. niger* after 7 days of incubation at 29 °C, according to the modified DIN 53931 Standard Method

Fungus	Growth				Intensity of sporolation	
	A*		B*		A	B
C. globosum	4	(> 75)	[0]	(0)	+++	/
A. niger	5	(100)	1	(5)	+++	+

* A – untreated sample, B – sample coated with AG-RB finish.
Tomšič et al., 2009.

5.1.2. Passive Antimicrobial Activity of FAS and DMeDHEU Finishes

Structure of Cellulose Fibers Modified by FAS and DMeDHEU Finishes

The formation of a FAS nanocomposite network on the fiber surface, and the cross-linking of cellulose macromolecules with DMeDHEU in the amorphous regions of fibers (Schindler & Hauser, 2004) are schematically presented in Figure 12. The presence of FAS and DMeDHEU on the cellulose fibers was confirmed by FT-IR spectra (Figures 13 and 14) and XPS measurements (Figure 15) (Tomšič et al., 2007, 2008). Comparison of the spectra of untreated cotton fabric (Figure 14a), cotton fabric coated with FAS finish (Figure 14b), as well as FAS coating on Al substrate (Figure 13) revealed well resolved bands attributable to the stretching modes of the C–F groups (Church & Evans, 1995; Hoffmann et al., 1997; Hoque et sl., 2006, 2007; Jiang et al., 2005; Lenk, 1994; Monde et al., 1999; Rabolt et al., 1984) showing beside the band at 1245 cm^{-1} and the band at 1237 cm^{-1}, while the band at 1204 cm^{-1}, which was well expressed in the FAS coating on Al substrate, was blurred by the cellulose –OH bending (1200 cm^{-1}) in the case of the coated cotton fabric. Moreover, the bands attributed to the Si–O–Si modes of silsesquioxane species formed when condensation proceeds in the presence of a textile substrate, could also not be assessed on the coated cotton fabric since the very strong bands ascribed to the cotton blurred the detailed absorption in the region from 1150-900 cm^{-1} (Fir et al., 2007; Tomšič et al., 2008; Vince et al., 2006). These results were fully confirmed by XPS measurements (Figure 15), showing the appearance of a fluorine (F 1s) peak at 689 eV and silicon (Si 2p) peak at 102 eV in the cotton fabric coated with FAS finish, since the XPS spectra of untreated cotton fabric revealed only two characteristic peaks belonging to carbon (C 1s) at 289 eV and oxygen (O 1s) at 533 eV.

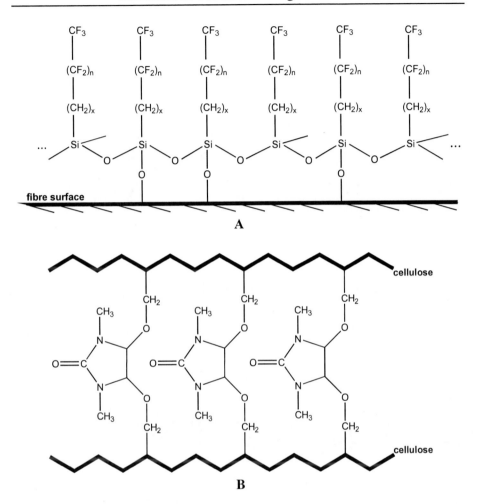

Figure 12. Formation of FAS nanocomposite network on the fiber surface (**A**) and cross-linking of cellulose macromolecules with DMeDHEU in the amorphous regions of fibers (**B**).

According to the cross-linking mechanism presented in Figure 12B, covalent bonds are formed between hydroxyl groups of DMeDHEU and cellulose molecules, in the condensation reaction during the curing stage of the finishing process (Schindler & Hauser, 2004). Since the finish molecules are small enough to enter the fibers, it was assumed that the cross-linking reaction occurs in the amorphous regions of the fibers. The incorporation of molecules of DMeDHEU into the cellulose structure was confirmed from the infrared spectral analysis. The ATR spectra (Figure 14c) revealed spectral changes in the $1800\text{-}1500$ cm^{-1} region.

For the finished cellulose, rather weak but characteristic bands attributed to the C=O and C–N stretching vibrations of DMeDHEU appeared at 1760, 1700 and 1260 cm^{-1} (Socrates, 2001). The broad absorption band at 1645 cm^{-1}, which is characteristic of the HOH bending vibrations of adsorbed water molecules (Kondo 1997; Łojewska et al. 2005; Hofstetter et al. 2006), occurred in the spectrum of untreated cellulose, while it was much weaker in the spectra of finished cellulose. Furthermore, the IR spectra of untreated and the treated cellulose samples showed a significant difference in the intensity and the shape of the band at 900 cm^{-1}, which is very sensitive to conformational changes of the interglucosidal bond (Mathlouthi and Koenig, 1986), proving the presence of cross-linking between DMeDHEU and cellulose macromolecules.

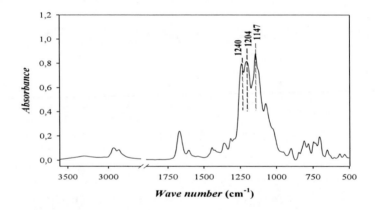

Figure 13. Measured ATR spectrum of FAS coating on Al substrate.

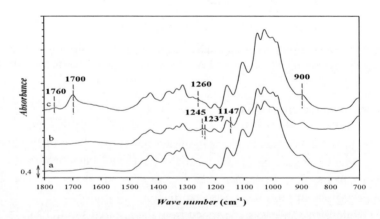

Figure 14. Measured ATR spectra of untreated cotton fabric (a) and of cotton fabric samples coated with FAS (b) and DMeDHEU (c) finishes.

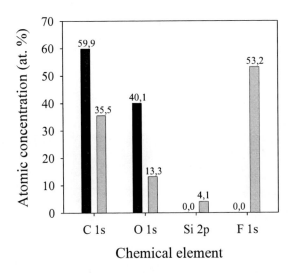

Figure 15. Surface composition of untreated cotton fabric (■) and cotton fabric coated with FAS finish (▓) obtained by XPS measurements.

The results presented in Figure 16 show that the application of FAS coating to cotton fabric highly increased the hydrophobicity of the untreated cotton fabric (Figure 16, IC and ID). On the other hand, the incorporation of DMeDHEU into the cellulose structure did not essentially decrease the hydrophilicity of the cellulose fibers (Simončič et al., 2008), which remained wettable by water (Figure 16, IB). It can also be seen from the results that water formed drops of high contact angles on the FAS coated cotton fabric samples even after 12 days of burial (Figure 16, IIC and IID) indicating that FAS polymer network was not attacked by microorganisms. The coating was still therefore present on the fibers, in spite of the fact, that the rotting process of coated cellulose macromolecules was in progress during the burial time, causing fiber damage with holes partially located in the fabric structure.

Biodegradation of Cellulose Fibers Modified by FAS and DMeDHEU Finishes

Comparison of the loss of breaking strength of cotton fabric during the burial time (Figure 8) revealed higher inhibition of the fibers rooting process in the presence of DMeDHEU than of FAS. This enables the conclusion that the mode of fiber protection of both finishes differs to each other and that the chemical modification of cellulose fibers with DMeDHEU provided more effective protection against microorganisms in the soil in comparison to that obtained with

FAS. These results were proved by the FT-IR spectra presented in Figure 17, where the band intensities at 1640 and 1548 cm^{-1}, characteristic for the Amide I and II, ascribed to the proteins produced during the growth of microorganisms on the fibers, are higher in the case of more degraded fibers.

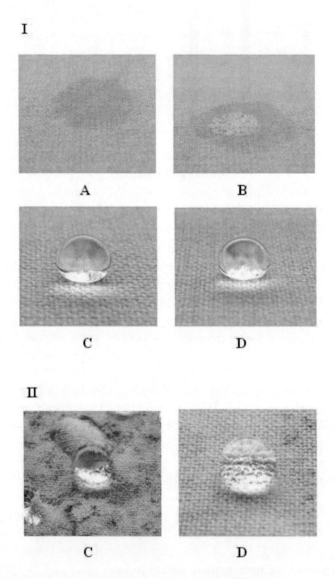

Figure 16. Photographs of water droplet placed on the untreated and finished cotton fabric samples before (I) and after (II) 12 days of burial. A) untreated sample, B) coated with DMeDHEU, C) coated with FAS, D) coated with FAS-DMeDHEU.

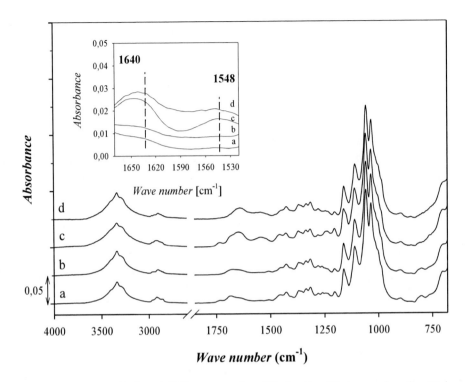

Figure 17. Measured infrared ATR spectra of modified cotton fabric samples after 12 days of burial. a) coated with FAS-DMeDHEU, b) coated with DMeDHEU, c) coated with FAS, d) untreated cotton fabric.

In order to discuss these results, we should consider both the binding mechanism of FAS and DMeDHEU to cellulose fibers as well as the mode of the microorganism action during the biodegradation process. It is clear that the formation of a FAS network of on the surface of cotton fabric (Figure 12A) caused an increase of the fiber hydrophobicity, as well as a high drop in their surface free energy, resulting in a decreased adhesion of microorganisms and impaired conditions of their growth. Namely, the microorganism growth is strongly influenced by different factors among which the presence of water is of great importance. A high decrease of fiber wettability thus significantly inhibited microbial degradation at initial burial time, but the protection was much less effective when the fibers were exposed to microorganism action for a longer time. In contrast to FAS, the application of DMeDHEU assured sufficient protection of fibers, in spite of the fact that they remained hydrophilic. The covalent bonds that formed between DMeDHEU and the cellulose macromolecules in the modification process (Figure 12B), strengthened the less ordered amorphous regions and held the macromolecular chains together, resulting in a decrease in cellulose chain mobility and fiber swelling. The powerful effect of the strengthening of amorphous region on inhibiting fiber biodegradation can also be explained by the manner of the cellulose biodegradation process, in which the action of the fungi is of great importance. Since fungi start to degrade fibers from their lumen outwards, the cross-linking of cellulose macromolecules inhibited the penetration of microorganisms into the amorphous region of the fibers where the biodegradation takes place. It can also be clearly seen from the results (Figure 8) that the FAS and DMeDHEU finishes act synergistically in the coating resulting in higher inhibition of the cellulose degradation process in comparison with single component FAS and DMeDHEU coatings. At the same time, microbiological tests according to the modified DIN 53931 and AATCC100-1999 standard methods revealed that neither DMeDHEU nor FAS provide active antifungal and antibacterial activity on the cellulose fibers.

Chapter 6

SOL-GEL FINISHES FOR PASSIVE
ANTIBACTERIAL ACTIVITY

6.1. SOL-GEL PRECURSORS AS MATERIAL OF CHOICE

It is clear that the antibacterial effect, closely related to the adhesion and colonization of bacteria on the cotton fabrics, is indirectly influenced by the low surface energy of the cotton fabrics. Zhao et al. (2004) demonstrated that by using a graded nickel-perfluoroethylene composite coating technique, it is possible to tailor the surface energy of membrane diffusers to optimum values, resulting in stable membrane filtration achieved by the retention of membrane permeability, which usually decreases due to the growth of a biofilm. It has been shown that the low dispersive part of the surface free energy in particular has a tremendous effect on reducing the attachment of *E. coli*. Like membrane diffusers, cotton fabric also has a porous structure and both substrates suffer problems of poor durability, lack of resistance to cleaning agents and are prone to leaching (Hamza et al., 1997). Chemical modification of cotton fabric with sol-gel based organic-inorganic hybrids containing perfluoroalkyl groups enables the formation of a nanocomposite coating of an organic-inorganic structure with extremely low surface free energy, providing excellent hydrophobicity and oleophobicity of fibers (Mahltig & Böttcher, 2003; Qing et al., 2002; Shao et al., 2004a, b; Tomšič et al., 2008; Yu et al., 2007). Mild processing conditions are required and single-step processing to impregnate the fabric is usually enough to impart multifunctional properties such as high contact angles for oil and water. Importantly, sol-gel hybrids exhibit excellent adhesion on cotton, attained through condensation between the -OH groups of the hydrolyzed silanes and those present at the surface of cellulose (Abdelmouleh et al., 2002; Vince et al., 2006). Since these studies have also shown that the sol-gel procedure enables the formation of textiles with radically new chemical and physical properties, which cannot not be achieved by the application of conventional chemical finishing agents, in addition to the FAS coating, novel sol-gel coatings were prepared with 1H, 1H, 2H, 2H-perfluorooctyltriethoxysilane (PFOTES) in combination with di-ureapropyltriethoxysilane [bis(aminopropyl)terminated-polydimethylsiloxane (1000) (PDMSU) (Vilčnik et al., 2009) and with aminopropyl-perfluoroisooctyl polyhedral oligomeric silsesquioxane ($AP_2PF_2IO_4$ POSS) (Figure 18) in combination with a commercial diisocyanatohexyl (DICH) cross-linker (Jerman et al., 2008). The letter represents a novel synthesised tri-functional POSS based silane precursor $R_xR'_yR''_z(SiO_{3/2})_8$, (x+y = 8) bearing di-aminopropyl, di-perfluorohexiletil and tetra-isooctyl groups. It has a well defined cage-like structure with a stable inorganic Si-O core of dimensions from 1 to 3 nm surrounded by organic (R) substituents, resembling in this respect organically functionalized nanosized particles of SiO_2. Most of the POSS known so far are not suitable for achieving low surface energy cotton finishes, because they do not prefer to form compliant gels as for example, when more common bridged silsesquioxanes (b-PS), prepared from different bi-podal alkoxysilane precursors (Shea & Loy, 2001) are used for cotton finishes. Stiff and structurally well defined polyhedral core of POSS prefer to form less compliant and more porous, water and oxygen permeable coatings and since usually they do not contain alkoxysilane reactive groups, the coatings could be washed from cotton fabrics and from other substrates relatively easy. In order to increase the fastness of POSS coatings on cotton fabrics and to overcome the absence of alkoxysilane groups in $AP_2PF_2IO_4$ POSS, well proven to enhance fastness of b-PS finishes on cotton fabrics (Vince et al., 2006), POSS polyhedra were functionalized with di-amiopropyl groups, known by their reactivity towards isocyanato groups. This enables the possibility to add an appropriate diisocyanato cross-linker to the corresponding POSS solution and afterwards applied on cotton fabric.

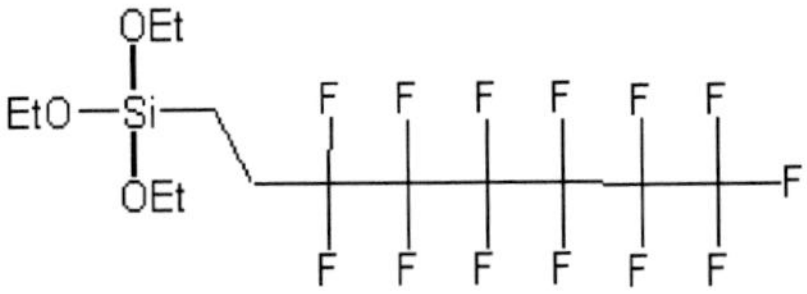

A) PFOTES

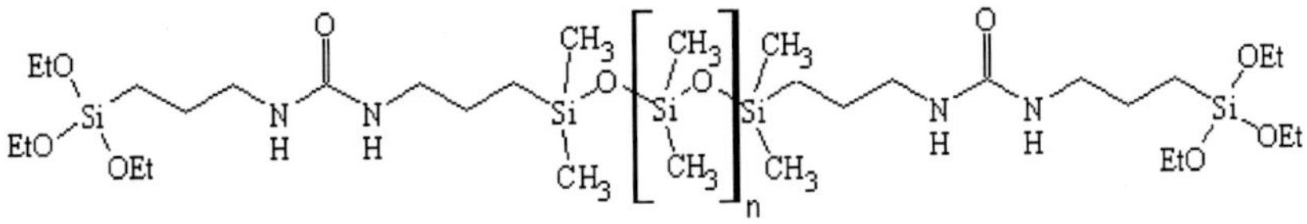

B) PDMSU

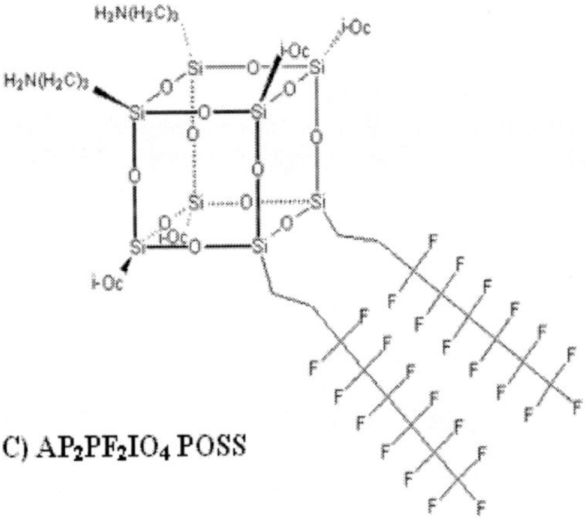

C) AP₂PF₂IO₄ POSS

Figure 18. Structure of 1H, 1H, 2H, 2H-perfluorooctyltriethoxysilane (PFOTES) (**A**), di-ureapropyltriethoxysilane [bis(aminopropyl)terminated-polydimethylsiloxane (1000) (PDMSU) (**B**) and aminopropyl-perfluoroisooctyl polyhedral oligomeric silsesquioxane ($AP_2PF_2IO_4$ POSS) (**C**).

6.2. STRUCTURE OF CELLULOSE FIBERS MODIFIED BY PFOTES-PDMSU AND AP₂PF₂IO₄ POSS FINISHES

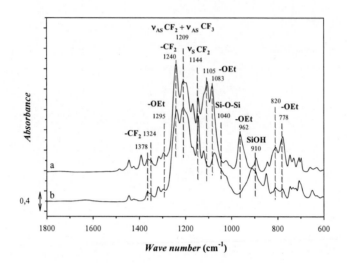

Figure 19. ATR spectra of non-hydrolyzed (a) and hydrolyzed (b) PFOTES deposited on an Al substrate.

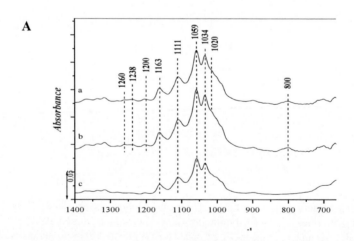

The structure of PFOTES-PDMSU and AP₂PF₂IO₄ POSS coatings was investigated by analyzing the IR ATR spectra. A closer look at the ATR spectra of the PFOTES coating deposited on an Al substrate (Figure 19) revealed the bands attributed to $v_a(CF_2)$ mixed with rocking (CF_2) (1240 cm^{-1}), $v_a(CF_2) + v_a(CF_3)$ (1209 cm^{-1}) and $v_s(CF_2)$ (1145 cm^{-1}) vibrations of the perfluoro chains (Lenk et al., 1994). The assignment of the C-F bands was reliable enough for performing subtractive spectral analysis, which we used for the identification of the C-F groups and how they changed with respect to the Si-CH$_3$ groups of the PFOTES-PDMSU finishes on cotton fabric. Inspection of the spectra in Figure 20A showed that the bands attributed to the PFOTES-PDMSU/cotton sample are dominated by bands belonging to the cotton itself. Subtraction of the IR spectrum of cotton from the IR spectra of the finished cotton (Figure 20B), however, showed bands attributed to PDMSU at 1256 cm^{-1} (δ(Si-CH$_3$)), 798 cm^{-1} (r(Si-CH$_3$)) and CF bands ($v_a(CF_2)$ mixed with rocking (CF_2) (1240 cm^{-1}), $v_a(CF_2) + v_a(CF_3)$ (1209 cm^{-1}) and $v_s(CF_2)$ (1145 cm^{-1})) modes. In addition, strong asymmetric and symmetric v(Si-O-Si) stretching bands of PDMSU chains at 1087 and 1020 cm^{-1} (Figure 20B) appeared, which can only be identified from the slight intensity changes of the main bands attributed to the cotton fabric in the region of 1200 - 1000 cm^{-1} (Chung et al., 2004; Langkilde & Svantesson, 1995). The bands of the urea-urea linkage were very weak, in contrast to the ATR spectra of cotton treated with PDMSU (Vince et al., 2006).

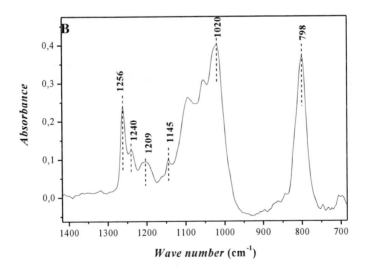

Figure 20. ATR IR spectra of cotton and cotton treated with PFOTES-PDMSU (**A**) and difference IR ATR spectra of PFOTES-PDMSU coatings obtained by subtracting the spectrum of cotton from the spectra of PDMSU-PFOTES/cotton (**B**).

The ATR IR transmission spectra in Figure 21 confirmed the interactions between $AP_2PF_2IO_4$ POSS and DICH since when the condensation of POSS and DICH proceeded, the band at 2269 cm^{-1} belonging to the isocyanate groups (which was accompanied by the appearance of the bands of the urea groups (Amid I and Amide II bands) disappeared from the spectrum. Furthermore, the presence of the band at 1116 cm^{-1} confirmed the formation of cube-like structures of $AP_2PF_2IO_4$ POSS. As expected, the subtractive spectra (Figure 22) clearly demonstrated the presence of POSS on the cotton fabrics, inferred from the C-F bands (1240, 1210 and 1145 cm^{-1}) (Lenk et al., 1994) and the ($-SiO_{3/2}$) stretching band of the POSS polyhedra at 1116 cm^{-1}. When DICH was added to the POSS/EtOH solution and applied on the cotton fabrics, the corresponding IR ATR spectra revealed the formation of bands characteristic of the formation of urethane bonds (Amide I and II bands at 1680 and 1540 cm^{-1}), confirming the reactions between POSS and DICH. The similarity with the spectra shown in Figure 21 is striking, despite the presence of cotton fibers, which blurred the appearance of bands in the corresponding IR ATR spectra.

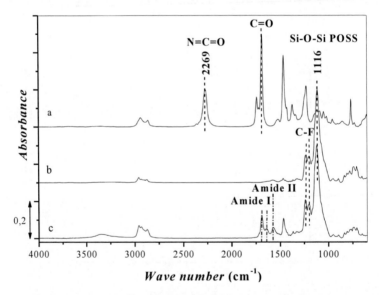

Figure 21. IR transmission spectrum of DICH (a), AP_2PFIO_4 POSS (b) and a mixture of DICH and AP_2PFIO_4 POSS (c) deposited on an Al substrate.

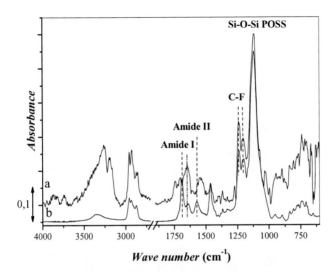

Figure 22. Difference IR transmission spectrum obtained by subtracting the spectrum of untreated cotton from the spectrum of AP$_2$PF$_2$IO$_4$ POSS/cotton (a) and IR transmission spectrum of DICH + AP$_2$PF$_2$IO$_4$ POSS deposited on an Al substrate (b).

6.3. HYDROPHOBICITY AND OLEOPHOBICITY OF THE SOL-GEL MODIFIED FIBERS

The results presented in Figure 23 showed that the presence of perfuoroalkyl groups in all three sol-gel hybrids, i.e., FAS, PFOTES and AP$_2$PF$_2$IO$_4$ POSS, provided the hydrophobicity and oleophobicity of the coatings deposited on Al substrates, resulting in high contact angles for water, formamide and diiodimethane. The highest contact angles were obtained for the FAS coating, followed by the PFOTES-PDMSU and AP$_2$PF$_2$IO$_4$ POSS coatings. The surface free energy values (Table 2) derived from the Van Oss equation (5) revealed that all three coatings represent highly apolar low energy surfaces with γ_S^{TOT} equal to 10.3 mJ/m^2 for FAS, 14.5 mJ/m^2 for PFOTES-PDMSU and 20.58 mJ/m^2 for AP$_2$PF$_2$IO$_4$ POSS. Concurrently, the polar electron-donor (γ_S^-) and electron-acceptor (γ_S^+) interactions decreased, reaching values lower than 0.8 mJ/m^2 and

0.09 mJ/m^2, respectively. The polarity of the coatings obtained from the Van Oss relation was very low, equal from 2 to 5 %.

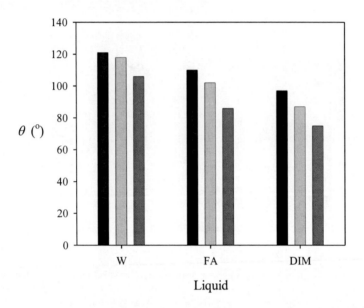

Figure 23. Contact angles of water (W), formamide (FA) and diiodomethane (DIM) on the Al substrates coated with FAS (■), PFOTES-PDMSU (▨) and AP$_2$PF$_2$IO$_4$ POSS (▨).

Table 2. The surface free energy components of the coatings on Al substrates obtained with the use of Equation (5) according to the Van Oss and co-workers approach

Coating	γ_S^{LW} (mJ/m^2)	γ_S^+ (mJ/m^2)	γ_S^- (mJ/m^2)	γ_S^{TOT} (mJ/m^2)	Polarity (%)
FAS	9.97	0.08	0.79	10.30	5
PFOTES-PDMSU	14.06	0.04	0.32	14.30	2
AP$_2$PF$_2$IO$_4$ POSS	20.12	0.08	0.61	20.58	2

The chemical modification of the cellulose fibers treated with the sol-gel hybrids was washing resistant since its water repellent properties (Figure 24) were retained after consecutive washing, as reflected in the water contact angles, which changed from 151° (unwashed) to 141° (10 x washed) for AP$_2$PF$_2$IO$_4$ POSS and from 147° (unwashed) to 142° (10 x washed) for PFOTES-PDMSU. Only FAS provided worse washing fastness since the initially high contact angles (149°) changed to 127° after 10 washings. It is important to note that the washing fastness of the coating is of great importance for textile products when in use.

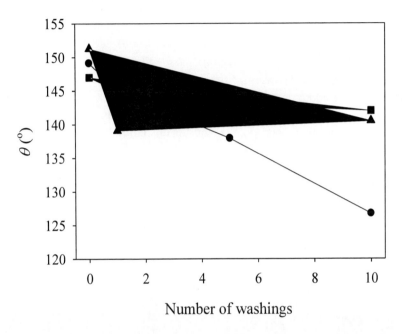

Figure 24. Static contact angle, θ, of water on finished cotton samples before and after repetitive washing at 40 °C. Coating: -●- FAS, -■- PFOTES-PDMSU, -▲- AP$_2$PF$_2$IO$_4$ POSS.

The results presented in Figure 24 show that, due to the rough morphology of woven fabrics (Tuteja, 2007), the contact angles for water increased from 121, 118 and 113°, observed on Al substrates, to 149, 147 and 151° for FAS, PFOTES-PDMSU and AP$_2$PF$_2$IO$_4$ POSS. On the basis of the high water contact angles, the work of adhesion values (Figure 25), which were calculated from the left side of equation (5), were very low because of the decrease of the solid surface free energy of the coated cotton fabric, indicating that there are very weak adhesion forces between water and the cotton fibers. The SEM micrographs (Figure 26) and AFM measurements (Figure 27) of uncoated and coated cotton fibers showed that both fibers had very similar rms values (~ 30 – 50 nm). This means that the texture of the cotton fabric, and not the roughness of the individual fibers, played the decisive role in the observed enhancement of hydrophobicity. This was attributed to air, which is trapped in the fiber assembly in the rough fabric surface. Namely, we believe that in addition to the chemical structure, the surface porosity, geometry and roughness can also influence the liquid contact angles (Černe et al., 2008; Yu et al., 2007). The observed apparent contact angles on unwashed cotton fabrics were similar to those observed for cotton treated with a perfluorooctylated quaternary ammonium silane coupling agent (Isquith et al., 1972) or hexadecyltrimethoxysilane (Mahltig et al., 2004), but higher than those obtained for PDMSU treated cotton fabrics (Vince et al., 2006) and tetrafluoroethylene copolymers (Phani, 2006), indicating that the cotton fabrics became superhydrophobic. A comparison of the results in Figures 20 and 21 shows that, despite the lowest liquid contact angles observed on Al substrates, the $AP_2PF_2IO_4$ POSS coating provided an excellent repellency of cotton fabric with a water contact angle higher than 150°. The reason for this could be the unique cube-like structure of the $AP_2PF_2IO_4$ POSS coating on the surface of the cotton fabric (Figure 28).

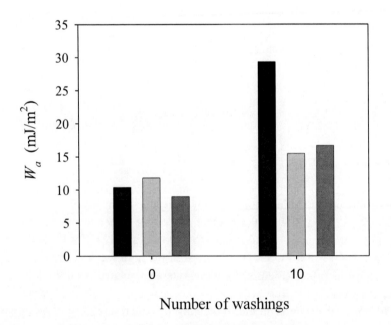

Figure 25. Work of adhesion, W_a of the cotton fabrics coated with FAS (■), PFOTES-PDMSU (■) and $AP_2PF_2IO_4$ POSS (■) before and after 10 washings.

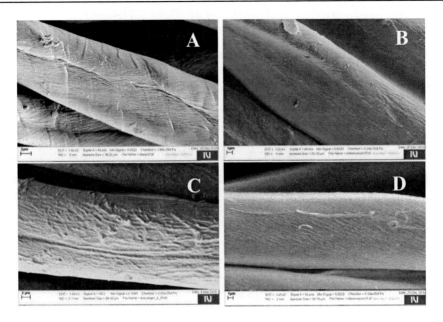

Figure 26. SEM micrographs of uncoated cotton fibers (**A**) and fibers coated with PFOTES-PDMSU (**B**), $AP_2PF_2IO_4$ POSS (**C**) and FAS (**D**) finishes.

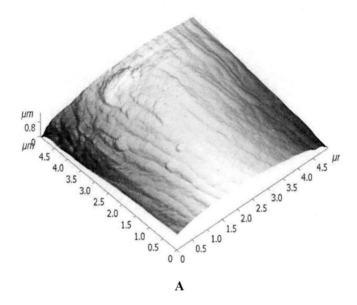

A

Figure 27. (Continued)

B

Figure 27. AFM of uncoated (**A**) and PFOTES-PDMSU coated (**B**) cotton fibers.

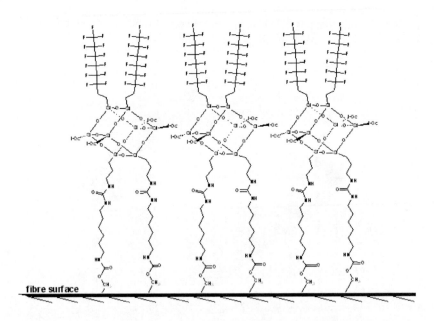

fibre surface

Figure 28. Schematic presentation of the $AP_2PF_2IO_4$ POSS coating formation and its orientation on the surface of cotton fabric.

6.4. THE INFLUENCE OF LOW SURFACE FREE ENERGY OF THE MODIFIED COTTON FIBERS ON THE BACTERIAL REDUCTION

Table 3. Reduction, R_A, of bacteria $E. coli$ (ATCC 25922) according to the EN ISO 20743:2007 Transfer Method determined on finished cotton samples before and after 1 and 10 cycles of washing

Coating	R_A (%)		
	0^a	1^a	10^a
Untreated	$/^b$	$/^b$	$/^b$
AG-RB	100	88.9 ± 3.4	$/^b$
FAS	99.3 ± 0.1	-	28.9 ± 9.9
PFOTES-PDMSU	100	$99.7 \pm 0,1$	60.6 ± 10.8

[a] 0 – unwashed sample, 1 – sample after the first washing cycle, 10 – sample after the tenth washing cycle;
[b] /: no reduction of bacteria.

In order to evaluate the low surface energy effect on the antibacterial properties and to distinguish it from the inherent antibacterial effect of the Ag particles, an alternative procedure for the assessment of bacterial growth was applied. Namely, the main problem encountered with the AATCC 100-1999 standard method is the need for complete wetting of the cotton fabrics when the antibacterial properties have to be assessed. We avoided this problem by using the EN ISO 20743:2007 Transfer Method, which gives the possibility of establishing, at least qualitatively, the contribution of the hydrophobicity to the antibacterial properties achieved. Inspection of the results shown in Table 3 shows that the antibacterial properties could be obtained with the application of an antibacterial active agent such as AgCl in an AG-RB coating, but also by applying hydrophobic and oleophobic FAS and PFOTES-PDMSU finishes to the fabrics. It seems that the presence of FAS and PFOTES-PDMSU prevents, or at least hinders, the adhesion of bacteria and their consequent growth and the formation of a biofilm on the finished fabrics. This effect, which we called "passive antibacterial activity", exhibited bacteria reduction even after 10 consecutive washings, suggesting the effective prevention of bacterial adhesion on low surface energy cotton fabrics. While no reduction of the growth of bacteria was observed for cotton coated with AG-RB after 10 repetitive washings, because of the gradual leaching of Ag cations from the coating the beneficial and long-lasting low surface energy effect of FAS and PFOTES-PDMSU in the absence of Ag particles was established, with a reduction in bacteria of about 28.9 ± 9.9 % for FAS and 60.6 ± 10.8 % for PFOTES-PDMSU. The latter results are in accordance with the washing fastness of the coatings, which was much higher in the case of PFOTES-PDMSU than with FAS.

CONCLUSION

In this work we have presented the results of the chemical modification of cellulose fibres aimed at protecting the textile material against biodegradation. Namely, cellulose fibres are highly susceptible to microbial attack, resulting in worsened technological and applicable properties of textile products. This is a particularly crucial problem for textiles that are in use. The rate and degree of cellulose biodegradation is affected by several factors, among which the most important are the genera of microorganisms and the environmental conditions needed for microbial growth. In view of this, sufficient antimicrobial protection of cellulose fibers can be obtained in two ways. The first involves chemical modification of fibers with an active antimicrobial agent, e.g. biocide or biostat, which are able to kill or inhibit the growth of bacteria and fungi. This active antimicrobial activity was demonstrated with the use of a finish on a base of AgCl in combination with an organic-inorganic binder. The other possibility involves chemical modification with finishes which are not antimicrobial agents but their presence on cellulose fibers significantly impedes conditions for microbial growth and therefore improves fiber resistance against microbial attack. This kind of antimicrobial protection, which was denoted 'passive antibacterial activity', was demonstrated by the application of easy-care and water and oil repellent finishes to cellulose fibers. While the easy-care finish strengthened the less ordered amorphous regions and therefore inhibited the penetration of microorganisms into the fibers, where biodegradation takes place, the water and oil repellent finish strongly decreased the fiber surface free energy, resulting in a highly decreased adhesion of microorganisms on the hydrophobic fibers.

ACKNOWLEDGMENTS

This work was supported by the Slovenian Research Agency (Programmes P2-0213, P1-0030 and Project M2-0104) and the Slovenian Ministry of Defense (Project M2-0104).

REFERENCES

Abdelmouleh, M.; Boufi, S.; ben Salah, A., Belgacem, M. N. & Gandini, A. (2002). Interaction of silane coupling agents with cellulose, *Langmuir, 18,* 3203–3208.

Akkopru, B. & Durucan, C. (2007). Preparation and microstructure of sol-gel derived silver-doped silica. *Journal of Sol-Gel Science and Technology, 43,* 227–236.

Bajaj, P. (2002). Finishing of tekstile materials. *Journal of Applied Polymer Science, 83,* 631–659.

Blackburn R. S. (Ed.). (2005). *Biodegradable and Sustainable Fibres.* Cambridge: Woodhead Publishing, Boca Raton: CRC.

Buschle – Diller, G., Zeronian, S. H., Pan, N. & Yoon, Y. (1994). Enzymatic hydrolysis of cotton, linen, ramie and viscose rayon fabrics. *Textile Research Journal, 64,* 270–279.

Clarke A. J. (1997). *Biodegradation of cellulose: enzymology and biotechnology.* Lancaster: Technomic publishing co.

Cao, Y. & Tan, H. (2002). Effects of cellulase on the modification of cellulose. *Carbohydrate Research, 337,* 1291–1296.

Cao, Y. & Tan, H. (2004). Structural characterization of cellulose with enzymatic treatment. *Journal of Molecular Structure, 705,* 189–193.

Cao, Y. & Tan, H. (2005). Study on crystal structures of enzyme-hydrolyzed cellulosic materials by X-ray diffraction. *Enzyme and Microbial Technology, 36,* 314–317.

Chung, C., Lee, M. & Choe, E. K. (2004). Characterization of cotton fabric scouring by FT-IR ATR spectroscopy. *Carbohydrate Polymers, 58,* 417–420.

Church, J. S. & Evans, D. J. (1995). The quantitative-analysis of fluorocarbon polymer finishes on wool by FT-IR spectroscopy. *Journal of Applied Polymer Science*, *57*, 1585–1594.

Colthup, N. B., Daly, L. H. & Wiberley, S. E. (1990). *Introduction to Infrared and Raman Spectroscopy*. Third Edition. San Diego: Academic Press.

Černe, L., Simončič, B. & Željko, M. (2008). The influence of repellent coatings on surface free energy of glass plate and cotton fabric. *Applied surface science*, *254*, 6467–6477.

Dring, I. (2003). Anti-microbial, rotproofing and hygiene finishes. In D. Heywood, (Ed.), *Textile Finishing* (pp. 351–371). Bradford: Society of Dyers and Colourists.

El-Morsy, E. M., El-Saht, H., El-Dohlob, S. M. & El-Metwally, M. M. (2001). Microfungi contaminated cotton-fabrics and their metabolic roles in biodeterioration. *Egyptian Journal of Biotechnology*, *10*, 179–197.

Evans, E. & McCarthy, B. (1998). Biodeterioration of natural fibres. *Journal of the Society of Dyers and Colorists*, *4*, 114–116.

Fir, M., Vince, J., Šurca Vuk, A., Vilčnik, A., Jovanovski, V., Mali, G., Orel, B. & Simončič, B. (2007). Functionalisation of cotton with hydrophobic urea/polydimethylsiloxane sol-gel hybrid. *Acta Chimica Slovenica*, *54*, 144–148.

Fischer, J. J., Morey, P. R. & Sasser, P. E. (1980). Gram-negative bacterial content and fiber properties of raw cotton. *Textile Research Journal*, *50*, 735–739.

Flemming, H. C. (1998). Relevance of biofilms for the biodeterioration of surfaces of polymeric materials. *Polymer Degradation and Stability*, *59*, 309–315.

Gao, Y. & Cranston, R. (2008). Recent advances in antimicrobial treatments of textiles. *Textile Research Journal*, *78*, 60–72.

Glazer, A. N. & Nikaido, H. (2001). *Microbial biotechnology: fundamentals of applied microbiology*. New York: W. H. Freeman and Company.

Goynes, W. R., Moreau, J. P., Delucca A. J. & Ingber, B. F. (1995). Biodeterioration of nonwoven fabrics. *Textile Research Journal*, *65*, 489–494.

Gupta, A., Maynes, M. & Silver, S. (1998). Effects of Halides on Plasmid-Mediated Silver Resistance in *Escherichia coli*. *Applied and Environmental Microbiology*, *64*, 5042–5045.

Hamza, A., Pham, V. A., Natsuura, T. & Santerre, J. P. (1997). Development of membranes with low surface energy to reduce the fouling in ultrafiltration applications. *Journal of Membrane Science*, *131*, 217–227.

Hoque, E., DeRose, J. A., Houriet, R., Hoffmann P. & Mathieu, H. J. (2007). Stable Perfluorosilane Self-Assembled Monolayers on Copper Oxide Surfaces: Evidence of Siloxy–Copper Bond Formation. *Chemistry of Materials, 19*, 798–804.

Hoque, E., DeRose, J. A., Hoffmann, P. & Mathieu, H. J. (2006). Robust perfluorosilanized copper surfaces. *Surface and Interface Analysis, 38*, 62–68.

Hoffmann, P. W., Stelzle, M. & Rabolt, J. F. (1997). Vapor Phase Self-Assembly of Fluorinated Monolayers on Silicon and Germanium Oxide. *Langmiur, 13*, 1877–1880.

Hulleman, S. H. D., van Hazendonk, J. M. & van Dam, J. E. G. (1994). Determinatin of crystallinity in native cellulose from higher plants with diffuse reflectance Fourier transform infrared spectroscopy. *Carbohydrate research, 261*, 163–172.

Isquith, A. J., Abbott, E. A., Walters P. A. (1972). Surface-bonded antimicrobial activity of an organosilicon quaternary ammonium chloride. *Applied Microbiology, 24*, 859–863.

Itävaara, M., Siika-Aho, M. & Viikari, L. (1999). Enzymatic degradation of cellulose based materials. *Journal of Environmental Polymer Degradation, 7*, 67–73.

Jerman, I., Šurca Vuk, A., Koželj, M., Orel, B. & Kovač, J. (2008). A structural and corrosion study of triethoxysilyl functionalized POSS coatings on AA 2024 alloy. *Langmuir, 24*, 5029–5037.

Jerman, I., Tomšič, B., Kovač, S., Simončič, B. & Orel, B (2008). Novel polyhedral oligomeric silsesquioxanes (POSS) as surface modifiers for cotton fabrics. In Z. Dragčević, (Ed.), *Magic world of textiles : book of proceedings* (pp. 370–375). Zagreb: Faculty of Textile Technology, University of Zagreb.

Jiang, W. C., Meng, W. D. & Qing, F. L. (2005). Synthesis of a novel perfluorooctylated polyacrylate and its application on cotton fabrics. *Journal of Applied Polymer Science, 98*, 222–226.

Jung, W. K., Kim, S. H., Koo, H. C., Shin, S., Kim, J. M., Park, Y. K., Hwang, S.Y., Yang, H. & Park, Y. H. (2007). Antifungal activity of the silver ion against contaminated fabric. *Mycoses, 50*, 265–269.

Kim, Y. H., Lee, D. K., Cha, G. C., Kim, C. W. & Young, S. K. (2007). Synthesis and Characterization of antibacterial Ag-SiO$_2$ nanocomposite. *Journal of Physical Chemistry, 111*, 3629–3635.

Kissa, E. (1984). Repellent finishes. In M. Lewin, & S. B. Sello, (Eds.), *Handbook of fiber science and technology: Volume II, Chemical procesing of fibers and fabrics: Functional finishes, Part B* (pp. 144–204). New York: Marcel Dekker.

Kusnetsov, J., Ivanainen, E., Nelomaa, N., Zacheus, O. & Martikainen, P. (2001). Copper and silver ions more effective against legionellae than against mycobacteria in a hospital warm water system. *Water Research, 35,* 4217–4225.

Langkilde, F. W. & Svantesson, A. J. (1995). Indentification of celluloses with Fourier-Transform (FT) mid-infrared, FT-Raman and near-infrared spectrometry. *Journal of Pharmaceutical & Biomedical Analysis, 13,* 409–414.

Lavrenčič Štangar, U., Orel, B., Vince, J., Jovanovski, V., Spreizer, H., Šurca Vuk, A. & Hočevar, S. J. (2005). Silicotungstic acid/organically modified silane proton-conducting membranes. *Journal of Solid State Electrochemistry, 9,* 106–113.

Lee, H. J. & Jeong S. H. (2005). Bacteriostasis and skin innoxiousness of nanosize silver colloids on textile fabrics. *Textile Research Journal, 75,* 551–556.

Lee, H. Y., Park, H. K., Lee, Y. M., Kim, K. & Park, S. B. (2007). A practical procedure for producing silver nanocoated fabric and its antibacterial evaluation for biomedical applications. *Chemical Communication, 28,* 2959–2961.

Lenk, T. J., Hallmark, V. M., Hoffmann, C. L., Rabolt, J. F., Castner, D. G., Erdelen, C. & Ringsdorf, H. (1994). Structural investigation of molecular-organization in self-assembled monolayers of a semifluorinated amidethiol. *Langmuir, 10,* 4610–4617.

Mahltig, B. & Böttcher, H. (2003). Modified Silica Sol Coatings for Water-Repellent Textiles. *Journal of Sol-Gel Science and Technology, 27,* 43–52.

Mahltig, B., Fiedler, D. & Böttcher, H. (2004). Antimicrobial sol-gel coatings. *Journal of Sol-Gel Science and Technology, 32,* 219–222.

Mahltig, B., Haufe, H. & Böttcher, H. (2005). Functionalisation of textiles by inorganic sol-gel coatings. *Journal of Materials Chemistry, 15,* 4385–4398.

Matsumura, Y., Yoshikata, K., Kunisaki, S. & Tsuchido, T. (2003). Made of bactericidal action of silver zeolite and its comparison with that of silver nitrate. *Applied Environmental Microbiology, 69,* 4278–4281.

Monde, T., Fukube, H., Nemoto, F., Yoko, T. & Konakahara, T. (1999). Preparation and surface properties of silica-gel coating films containing branched-polyfluoroalkylsilane. *Journal of Non-Crystalline Solids, 246,* 54–64.

Montegut, D., Indictor, N. & KOESTLER, R. J. (1991). Fungal deterioration of cellulosic textiles: a review. *International Biodeterioration & Biodegradation, 28,* 209–226.

Moulder, J. F., Stickle, W. F., Sobol, P. E. & Bomben, K. D. (1995). *Handbook of X-Ray Photoelectron Spectroscopy*. Eden Prairie, MN: Physical Electronics Inc.

Nelson, M. L. & O'Connor, R. T. (1964). Relation of Certain Infrared Bands to Cellulose Crystallinity and Crystal Lattice Type. Part II. A New Infrared Ratio for Estimation of Crystallinity in Celluloses I and II. *Journal of Applied Polymer Science, 8,* 1325–1341.

Park, C. H., Kang, Y. K., & Im, S. S. (2004). Biodegradability of cellulose fabrics. *Journal of Applied Polymer Science, 94,* 248–253.

Phani, A. R. (2006). Structural, morphological, wettability and thermal resistance properties of hydro-oleophobic thin films prepared by a wet chemical process. *Applied Surface Science, 253,* 1873–1881.

Qing, F. L., Ji, M., Lu, R., Yan, K. & Mao Z. (2002). Synthesis of perfluoroalkyl-containing multifunctional groups compounds for textile finishing. *Journal of Flourine Chemistry, 113,* 139–141.

Rabolt, J. F., Russell, T. P. & Twieeg, R. J. (1984). Structural studies of semifluorinated normal-alkanes. 1. synthesis and characterization of $F(CF_2)N(CH_2)MH$ in the solid-state. *Macromolecules, 17,* 2786–2794.

Raschle, P. (1989). Microbial influence on cellulosic textiles and microbiological testing. *International Biodeterioration, 25,* 237–244.

Schindler, W. D, & Hauser, P. J. (2004). *Chemical Finishing of Textiles.* Cambridge: CRC Press.

Shao, H., Meng, W. D. & Qing, F. L. (2004). Synthesis and surface antimicrobial activity of a novel perfluorooctylated quaternary ammonium silane coupling agent. Journal of Fluorine Chemistry, *125,* 721–724.

Shao H., Sun, J. Y. & Meng, W. D. (2004). Water and Oil Repellent and Durable Press Finishes for Cotton Based on a Perfluoroalkyl-Containing Multy-Epoxy Compounds and Citric Acid. *Textile Research Journal, 74,* 851–855.

Seventekin, N. & Ucarci, O. (1993). The damage caused by micro-organisms to cotton fabrics. *Journal of Textile Institute, 84,* 304–313.

Simončič, B. (2003). Importance of antimicrobial agents for textile finishing. *Tekstilec, 46,* 64–72.

Simončič, B., Černe, L., Tomšič, B. & Orel, B. (2008). Surface properties of cellulose modified by imidazolidinone. *Cellulose, 15,* 47–58.

Socrates G. (2001). *Infrared and Raman Characteristic Group Frequencies.* New York: John Wiley & Sons, LTD.

Szostak – Kotowa, J. (2004). Biodeterioration of textiles. *International Biodeterioration & Biodegradation, 53,* 156–170.

Tomšič, B., Simončič, B., Orel, B., Vilčnik, A. & Spreizer, H. (2007). Biodegradability of cellulose fabric modified by imidazolidinone. *Carbohydrate Polymers*, *69*, 478–488.

Tomšič, B., Simončič, B., Orel, B., Černe, L., Forte-Tavčer, P., Zorko, M., Jerman, I., Vilčnik, A. & Kovač, J. (2008a). Sol-gel coating of cellulose fibres with antimicrobial and repellent properties. *Journal of Sol-Gel Science and Technology*, *47*, 44–57.

Tomšič, B., Simončič, B., Žerjav, M. & Simončič, A. (2008b). A low nutrition medium improves the determination of fungicidal activity of AgCl on cellulose fibres. *Tekstilec*, *51*, 231–241.

Tomšič, B., Simončič, B., Orel, B., Žerjav, M., Schroers, H.-J. Simončič, A. & Samardžija, Z. (2009). Antimicrobial activity of AgCl embedded in a silica matrix on cotton fabric. *Carbohydrate. Polymers*, *75*, 618–626.

Tomšič, B., Simončič, B., Cvijin, D., Orel, B., Zorko, M. & Simončič, A. (2008). Elementary nano sized silver as antibacterial agent on cotton fabric. *Tekstilec*, *51*, 199–215.

Tuteja, A., Choi, W., Ma, M. L., Mabry, J. M., Mazzella, S. A., Rutledge, G. C., McKinley, G. H. & Cohen, R. E. (2007). Designing superoleophobic surfaces. *Science*, *318*, 1618–1622.

Van Oss, C. J., Chaudhury, M. K. & Good, R. J. (1988a). Interfacial Lifshitz-van der Waals and polar interactions in macroscopic system. *Chemical Review*, *88*, 927–941.

Van Oss, C. J., Good, R. J. & Chaudhury, M. K. (1988b). Additive and nonadditive surface tension components and the interpretation of contact angles. *Langmuir*, *4*, 884–891.

Vigo, T. L. (1983). Protection of textiles from biological attack. In S. B. Sello, (Ed), *Functional finishes, Part A, Chemical processing of fibres and fabrics, Handbook of fiber science and tehnology. Volume II* (pp. 367–426). New York and Basel: Marcel Dekker, Inc.

Vilčnik, A., Jerman, I., Šurca Vuk, A., Koželj, M., Orel, B., Tomšič, B., Simončič, B. & Kovač, J. (2009). Structural properties and antibacterial effects of hydrophobic and oleophobic sol-gel coatings for cotton fabrics. *Langmuir*, doi: 10.1021/la803742c.

Vince, J., Orel, B., Vilčnik, A., Fir, M., Surca Vuk, A., Jovanovski, V. & Simončič, B. (2006). Structural and Water-Repellent Properties of a Urea/Poly(dimethylsilixane) Sol-Gel Hybrid and Its Bonding to Cotton Fabric. *Langmuir*, *22*, 6489–6497.

Wang, J. X., Wen, L. X., Wang, Z. H. & Chen, J. F. (2006). Immobilization of silver on hollow silica nanospheres and nanotubes and their antimicrobial effects. *Materials Chemistry and Physics*, *96*, 90–97.

Xing, Y., Yang, X. & Dai, J. (2007). Antimicrobial finishing of cotton textile based on water glass by sol-gel method. *Journal of Sol-Gel Science and Technology*, *43*, 187–192.

Yu, M., Gu, G., Meng, W.-D. & Qing F.-L. (2007). Superhydrophobic cotton fabric coating based on a complex layer of silica nanoparticles and perfluorooctylated quaternary ammonium silane coupling agent. *Applied Surface Science*, *253*, 3669–3673.

Zhao, Q., Wang, S. & Muller-Steinhagen, H. (2004). Tailored surface free energy of membrane diffusers to minimize microbial adhesion. *Applied surface science*, *230*, 371–378.

INDEX